Routledge Revivals

The Anatomy of Literary Studies

First published in 1980, *The Anatomy of Literary Studies* provides students of English Literature with a clearer understanding of the significance and scope of the subject and a comprehensive background to its study. It gives pointers towards intellectual integrity and advice on independent study, libraries, essay writing and examinations. This reissue of Marjorie Boulton's classic work will be of particular value to students studying English at university or those applying to a course who would like a fuller understanding of what it might entail.

The Anatomy of Literary Studies

An Introduction to the Study of English Literature

Marjorie Boulton

Routledge
Taylor & Francis Group

First published in 1980
by Routledge & Kegan Paul Ltd

This edition first published in 2013 by Routledge
2 Park Square, Milton Park, Abingdon, Oxon, OX14 4RN

Simultaneously published in the USA and Canada
by Routledge
711 Third Avenue, New York, NY 10017

Routledge is an imprint of the Taylor & Francis Group, an informa business

Publisher's Note
The publisher has gone to great lengths to ensure the quality of this reprint but points out that some imperfections in the original copies may be apparent.

Disclaimer
The publisher has made every effort to trace copyright holders and welcomes correspondence from those they have been unable to contact.

A Library of Congress record exists under LC control number: 79041811

ISBN 13: 978-0-415-72241-4 (hbk)
ISBN 13: 978-1-315-85825-8 (ebk)
ISBN 13: 978-0-415-72249-0 (pbk)

THE ANATOMY OF LITERARY STUDIES

An introduction to the study of English Literature

MARJORIE BOULTON

Routledge & Kegan Paul
LONDON, BOSTON AND HENLEY

First published in 1980
by Routledge & Kegan Paul Ltd
39 Store Street,
London WC1E 7DD,
9 Park Street,
Boston, Mass. 02108, USA and
Broadway House,
Newtown Road,
Henley-on-Thames,
Oxon RG9 1EN

Set in 10/12pt Janson
and printed in Great Britain by
Lowe & Brydone Printers Ltd, Thetford, Norfolk

British Library Cataloguing in Publication Data

Boulton, Marjorie
The anatomy of literary studies.
1. English literature – Study and teaching
(Higher) – England
2. English literature – Study and teaching
I. Title
820'.7'1142 PR51.G7 79-41811

ISBN 0-7100 0441-9
ISBN 0-7100 0442-7 Pbk

à non moins docte et prudente,
que genereuse et vertueuse dame

Kathleen M. Hall

CONTENTS

PREFACE

Like Roger Ascham in his *Scholemaster*, 'in this litle booke I purpose to teach a yong scholer to go, not to daunce: to speake, not to sing . . . and after in good order and dew tyme to be brought forth to the singing and dauncing schole'; I have tried to write a simple, elementary and, I hope, sympathetically companionable guidebook for the student who is wondering whether to apply for a university course, or some other tertiary course, in English Language and Literature; or has been accepted for one; or has embarked on one. To some, it will all seem obvious; but I believe there are many others for whom it may fulfil a need. I hope I may help some students to achieve higher standards, and more satisfaction, in their studies; the two aims are not contradictory, but complementary.

This seems the place to mention two trifles. Whenever I refer to a hypothetical student as *he*, I refer to the species, not the sex; *he or she*, or such contrived suggestions as *hesh* or *hse*, over and over again, would be irritating; but I must make it clear that the implied *shes* are exactly as important as the explicit *hes*. Second, to avoid any appearance of unseemly complacency in my commendations of university teachers, let me add that I am not myself one, only grateful to many.

Finally, I must here express my specific gratitude to several scholars who have given me help in relation to this book: to Andrew Harvey, Daphne Hereward, Vincent Knowles, Sally Purcell, Humphrey and Julie Tonkin; to Georgina Warrilow and all the ever courteous and patient staff of the

Bodleian Library; and above all to the unwearyingly helpful friend to whom this book is dedicated.

M. B.
Oxford.

I

WHY STUDY LITERATURE?

There is a great deal of difference between the eager man who wants to read a book, and the tired man who wants a book to read.

G. K. Chesterton, *Charles Dickens*[1]

Ils sont là, hauts de cent coudées,
Christ en tête, Homère au milieu,
Tous les combattants des idées,
Tous les gladiateurs de Dieu.

Victor Hugo, *Les Mages*[2]

The most incontrovertibly useful function of the printed word is the storage and spread of information, as in a manual of navigation or nursing. A course in 'Literature' deals almost entirely with the more mysterious and ambiguous functions of the printed word: fictive creations of various kinds; attempts to communicate emotions and value judgments; devices of rhetoric, symbolism, imagery, evocation; wit, humour, fantasy, speculation: that vast and varied field which we may define as 'imaginative literature' as opposed to 'information', so long as we remember that we can have neither a perfect definition nor a rigidly exclusive delimitation of either.

It is mostly 'imaginative literature' that makes 'reading' one of the best pastimes known to man. Reading as a form of cheap, quiet entertainment has kept millions of people out of mischief and given them millions of hours of largely harmless pleasure: useful effects that may be achieved equally well by television, gardening or stamp collecting. A mere fondness for 'reading', while obviously a basic requisite for taking a degree course in English Literature (or any other adult course, as for instance in a college of education), no more fits anyone for entry to such a course than a good appetite qualifies anyone to embark upon a *Cordon Bleu* course in cookery.

Someone may hugely enjoy reading, yet have no interest whatever in literature as one of the great arts. Such a reader may not necessarily relish nothing above the level of thrillers, detective stories, westerns, 'romantic novels' in the commercial sense, newspapers and magazines. Some novels of acknowledged literary merit may be enjoyed: *Jane Eyre*, *Pride and Prejudice*, *Great Expectations*, *Heart of Midlothian*; possibly some Shakespeare – perhaps *Romeo and Juliet*, probably not *Timon of Athens* or *Measure for Measure*; possibly some poetry – *The Highwayman*, *Lays of Ancient Rome*, even *The Rime of the Ancient Mariner*; hardly Donne or Milton, Dryden or Auden; Wordsworth's rainbow and Tennyson's bugle, but hardly *The Prelude* or the whole of *In Memoriam*.

This innocent reader is soothed or thrilled probably most of all by *stories* – sequences of events with suspense and surprises; then probably by characters with whom he can in some way relate emotionally; by bits of vicarious living and gratifying fantasy; by happy expressions of his own sentiments, or noble, comforting, encouraging thoughts about life; even by ignoble thoughts that he finds comforting. He wants excitement, emotions, amusement, consolation. For him books, whether the shoddiest trash or the most complex productions of subtle art, minister to fairly obvious and perhaps almost universal psychological wants.

There are many far worse ways of passing the time. Some civilizing influences often rub off on the habitual lazy escapist reader: a stock of miscellaneous information; a little tolerance, a few broader sympathies and maturer insights; an improved articulateness; at least some possibility of conversation going beyond grievances and greeds. We see more by gazing idly through the window than by staring at the wall.

However, a degree course in English cannot be an opportunity for three years of cosy, relaxed, desultory reading in which the reader is almost passive. No one can be happy in any British university course unless he is willing to read a great deal that no one would read purely for relaxation; to study literary techniques and make or discuss critical appraisals; to consider books in relation to their historical contexts; to extend his mind to see books in the perspective of a time-scale covering thirteen centuries and also to concentrate his attention at times on the minutely close examination of single works, single pages, sometimes even single words.

Personal responses will remain important; to parrot some critic with no attempt to arrive at one's own defensible opinion of a book is insincere as well as dishonest; but personal responses on the level of 'I like that!' or 'This is a bore!', however sincere, will be useless. Raw initial reactions will often have to be corrected by more knowledge, closer analysis, discussion, even processes of personal maturing. Mental activity on this level is interesting, probably inexhaustibly so; easy or cosy it is not.

Furthermore, any single-subject honours course in English will require considerable study not only of 'Literature' but of 'Language'. In rather more than half the British universities this includes some Old English, which is more remote from our previous linguistic experience than Latin and has to be learned rapidly, as a foreign language, without the elaborate aids and plentiful jam on the pill normally used in teaching, say, French at school. In at least 75 per cent of British universities Middle English as such is obligatory; not all of it is as immediately rewarding as Chaucer. Some of the alternatives may be even more demanding; anything, for instance, called 'Linguistics', or 'Stylistics', or 'History of Language', may often prove as unfamiliar and exacting to many students as Old English would have been. So the total programme will at times seem daunting even to someone who passionately wanted to read English.

Therefore, someone who does not in any sense passionately want to read English – though such a passion will usually be patchy and fluctuating, as all our loves are imperfect – is wiser not to apply for such a course. A degree course in English is intended to be just as rigorous, in its way, as a course in mathematics or philosophy: a matter of ratiocination, system, discipline.

In a world in which a majority are underfed, various obscurantisms are enforced by sophisticated techniques of power and multitudes remain debarred from bare literacy, it is an extraordinary privilege to have some years of guided advanced study, with a modicum of comfort and privacy, where free enquiry is taken for granted. Few things can be better for society than a thick layer of people who have enjoyed a training at least aimed at inculcating intellectual honesty, exactness, mature sensibility and some awareness of the multiplicity, elusiveness and importance of truth. But such a training should have a severity to match its luxury.

The boy or girl thinking of reading English should also realize that a degree certificate will not be an oyster-knife for opening the world. Some arts students graduate into unemployment. There is fierce competition for entry to relevant professions felt by many to be alluring: the stage, publishing, broadcasting, journalism, and the entrant is likely to have a long slog before coming within sight of the big part, the signed column or the exciting responsibility. Devoted students prepared to take dons' modest salaries for the happiness of continuing in the studies they love may also be disappointed in the competition. School-teaching can be corroding misery for precisely those graduates who most truly love literature, an unremitting exposure to the brutalities of immature philistines, unless the love of learning is supplemented by an unsentimental love for even aggressively unwilling learners that not all of us can achieve. Many graduates in English have eventually to do work not much related to their field of study. Paradoxically, this is an additional reason why the course should be rigorous: a tough course demanding versatility, adaptability, perseverance and hard thinking should produce a mind pretty well trained for mastering other reasonably kindred skills; a too cosy course would produce a mushier, less adaptable mind.

Whether a course in English is the best vocational training for the aspiring serious writer is uncertain. We do not yet have massive evidence, since it was not possible to take a degree in English until nearly the end of the nineteenth century; until then most university-trained writers had studied classics. However, writers who graduated in English have already included: Kingsley Amis, W. H. Auden, William Golding, Aldous Huxley, Elizabeth Jennings, Philip Larkin, C. S. Lewis, Malcolm Lowry, V. S. Naipaul and John Wain; Robert Graves, with a research degree; William Empson and J. B. Priestley with degrees partly in English; so such studies evidently need not frighten the Muse away; and a lifelong dedication to English scholarship did not prevent J. R. R. Tolkien from writing what was not only a most unusual work of imaginative fiction, but has already sold over three million copies.

On the other hand, the writer with no university education whatever need not worry unduly in the company of Blake, the Brontës, Bunyan, Burns, Conrad, Dickens, George Eliot, Hardy, Edwin Muir, Pope, Shaw, Yeats – and Shakespeare. Obviously the close study of literature should teach a young author much about technique and standards. It might over-encourage imitation, but that can be a useful stage in development. It might inhibit a writer by reminders of how much has already been written, or by the pressure of achievements he cannot hope to equal; but an uninformed complacency about our own inferior work is not an unmixed blessing.

Because there is some element of luxury in English Studies in a world full of crying needs, and because such studies are not as clearly vocational as, say, engineering, the student of English is sometimes challenged to defend his choice. The vulgarly aggressive enquirer merely making dismissive assertions about 'your useless subject' has no more right to a courteous, considered answer than has any other rude oaf who wants a wince, not an answer, such as the one who sneers at someone's sweetheart, home, or religion; though it may do him good to receive an answer as gentle and reasoned as his question was offensive. Yet, in fairness, there is a genuine question that can be asked with decency. Why is the study of literature worth while?

There is no real consensus on this. Different religious or philosophical commitments, different varieties of subjective experience, will inevitably produce varying points of view. Not everyone even takes it for granted, as I do, that the arts are one of the things that make life itself meaningful and worth having, that give the human species its awful and ambiguous uniqueness. Bigots have rejected culture as worldly; barbarians have scorned it; and totalitarians try to cast it in rigid moulds and make projectiles of it.

We may reasonably hope that a course of careful attention to linguistic techniques will tend to counter the corruptions of language that are among our moral corrosions: our own deceptions and self-deceptions; the tricks of politicians, agitators, advertisers, assorted axe-grinders; those slovenly expressions of imprecise thinking that fill our lives with untruths and insincerities when we are not consciously willing any falsehood. We may hope, too, that the content of good literature is for the most part a good influence.

Unhappily, we cannot be as confident about the civilizing powers of literature as we could wish. William Joyce, hanged in 1946 as a British traitor after a peculiarly odious career of lying, bullying, gloating broadcasts, and recruiting more excusable traitors from prisoners-of-war under pressure, in the service of the Nazi German government, took first-class honours in English at London University.[3] How anyone, knowing some of the things Joyce must have known to achieve this, could ally himself wilfully with some of the most unmitigated wickedness in history, leaves us wailing with Lear, 'Is there any cause in nature that makes these hard hearts?'[4] We who love literature know the philistines are wrong, but a moral disaster so total warns us not to be smug; and a glance into our own hearts should set off the same warning, though not so clangorously. It is fair to say that Joyce seems to have shown his best self in relation to literary studies: as a teacher he was not only capable, but kind and patient.[5]

Any definition I can offer of the function and value of literature can be only elementary, eclectic and tentative; no more than a basis for discussion, a preliminary sketch-map for anyone who wishes to start exploring.

Literature gives us four Rs.

I RECREATION

Poetry may have had remote origins partly in magic, prose in annals, or drama in religious ritual; but for centuries recreation has been the most general and obvious function of imaginative literature. It has brought variety, interest, excitement, to drab lives; assisted consolatory fantasies; distracted people from their troubles; made them laugh. A play may give purpose to an evening out, a novel give savour to an evening in; we enjoy the experiences, and may later enjoy chatting about them. Shakespeare put so much of a broad, deep and compassionate mind into his plays, that they are virtually inexhaustible; but they were, in the first place, entertainments.

Mere entertainment, relaxation, amusement, are important, not only because pleasure is desirable unless it is an obstacle to something more important, but because most of us, without some unwinding mechanism, would go mad; most of us become at least unreasonable under stress alarmingly soon.

If literature were not in part entertainment, there could be no literary trade. Commercialism may, obviously, lower standards; it may pay to produce volumes of pappy sentimentality or over-seasoned sensationalism; but if there had not been monetary rewards for writers, we should probably not have the works of Shakespeare, Johnson, Dickens, Trollope, Tennyson, Conrad, Hardy, even perhaps Eliot or Auden. It is largely because of the recreational function of literature that books can be widely distributed and authors rewarded.

Recreation is the bait that first attracts us to reading. Professor of Literature D. J. Enright reminds us:[6]

> that in the Twenties
> and later our staple diet was *Red Letter*
> And Ethel M. Dell and *Old Moore's Almanac*,
>
> And that if you can escape for a moment
> And a moment's escape is all you can manage,
> No-one has the right to forbid you.

Many readers never go beyond recreation. The student must.

2 RECOGNITION

'How true!' is a relishing remark. Much of our enjoyment of literature comes from recognizing, like Pope, 'What oft was thought, but ne'er so well express'd . . .'.[7] One reason why we can greatly enjoy *Othello*, although the story is almost unrelievedly painful, is of course the poetry – musical, evocative, rich in images; but we also find a kind of pleasure in recognizing the truth of these fictive representations when we compare them with everyday observation (and newspaper reading): what the dreadful destructiveness of sexual jealousy can do to a man of tenderness, dignity, courage and intelligence; how the most sweet and generous love may fail if it confronts something too far outside its understanding; how plausible and even pleasant a destructive cynic can be when he presents his fake realism as helpful worldly advice to the inexperienced; how the sight of happiness, success and excellence arouses in some people the craving to spoil and poison. Unhappy human relations of the less spectacular kind are nearly as painful, but in *Middlemarch* we enjoy the skilled portrayal of misunderstandings and incompatibilities between Dorothea and Casaubon, Rosamond and Lydgate and others; the awkward conversations are brilliantly authentic and we admire the truth to experience. Loquacious, digressive rambling bores us in real life, but we love Miss Bates in *Emma* precisely because we still meet people who talk just as she does.

We enjoy, too, recognizing the truth of little details:

> And icicles, that fret at noon,
> Will eke their icy tails at night
> Beneath the chilly stars and moon.
>
> John Clare, *February*[8]

The blue way of the canal wound softly between the autumn hedges, on towards the greenness of a small hill. On

> the left was the whole black agitation of colliery and railway and the town which rose on its hill, the church tower topping all. The round white dot of the clock on the tower was distinct in the evening light.
>
> D. H. Lawrence, *The Rainbow*[9]

Such aesthetic satisfaction as we get from the better type of detective story comes, obviously, in part from the well constructed plot; but I think some comes from a similar recognition of truth in particulars:

> One or two of the pieces fitted firmly into place, but so many wouldn't fit at all. It was like doing the light-blue sky at the top of a jigsaw, with no clouds, not even a solitary sea-gull to break the boundless monochrome.
>
> Colin Dexter, *Last Seen Wearing*[10]

Often this pleasure of conscious recognition is a beginning of literary criticism; when we relish the exact conveying of some truth, trivial or tremendous, we are, even if unaware of that as such, appreciating style. When we instantly savour the perfection of

> Arthur Clennam came to a squeezed house . . .[11]

we feel how that *squeezed* for the house in Mews Street catches at once an appearance, an economic status and a moral atmosphere, and we admire genius for choosing a word so concentratedly apt.

3 REVELATION

When we *recognize* in literature what we already know from experience, we do not merely receive an identical echo of our experience. As John Wain puts it, 'we are seeing our experience through the lens of another mind, offering us its perceptions and its ordering pattern . . .'[12] – a new lens, not just a mirror. Reading George Eliot, I instantly *recognize* some of my own mechanisms of selfish self-deception; but I *see* them with a new clarity that helps me to combat them better.

However, if we feel a need to find moral justifications for literature, we shall probably find them mostly in the function I call Revelation: literature extends our experience.

We enjoy scraps of new information: Ian Fleming's details of what the affluent and knowing may eat abroad; Ben Traven's horrifying details of stokehold work in *The Death Ship*. More important, through books we can gain some slight notion of experiences far removed from our own: the comfortable learn something of poverty and its effects; someone of one race, of the culture of another; the sedentary, something of the life of action; the comparatively free, of the extra miseries of life under any totalitarian régime. Rather rational, cerebral people can learn more about the primal passions; all of us can travel in time as well as space; we may learn a little of the special motives of an artist, an inventor, an explorer; even of a saint, a criminal or a madman. The celibate or childless may learn a little about sex, marriage and parenthood. Men may learn much about women's experience of life, and women about men's.

Nothing in even the greatest literature can be regarded as wholly reliable; all writers have in some measure all the handicaps the rest of us have: no one can make perfectly accurate observations about everything; we all have our prejudices; and our powers of expression probably always fall short of our experience. Notably, males writing about females and females writing about males have inevitable limitations, though the greatest writers come near to transcending even these.

Someone might argue, not absurdly, that obvious fantasy does not enlarge our experience of life: that the fictions of Wells or Asimov, Tolkien or Ballard, Lewis Carroll, Alan

Garner or Richard Adams, are not even intended to portray reality. But the best fantasy does not function only as recreation; the human figures often exemplify credible human reactions to novelty, crisis or disaster; fantasy often extrapolates from real trends or problems to make us think; and all that man can imagine is a part of man's experience. There is a common-sense realism that knows milk comes in pint bottles in an electric van; but Coleridge's awe before one who had drunk the milk of paradise is not meaningless. We have dreams, imaginings, occasional hallucinations; we make myths, utopias, religions, superstitions, seasonal rites, institutional rituals. Good fantasy literature may indeed offer to some readers very valuable revelations – psychedelic in the best sense, without treacherous drugs.

The revelatory effects of wide reading should do something to broaden our sympathies, widen our tolerance and dissuade us from those boorish, brutal generalizations that demand conformity to norms too narrow, to supposedly universal criteria that in fact are only those of our own little circle. Even if we have never left our own village, to have seen a terrestrial globe teaches us something about scale.

4 REDEMPTION

A terrestrial globe is very instructive; but sometimes we may be allowed a sight of a celestial globe.

In using the term *redemption* for one function of literature I am certainly not intending to make a theological statement or suggestion; but I am also not just alliterating. I think there are moments in literature that have something in common with some religious experiences. They are infrequent and unforgettable. They offer glimpses of our undeveloped, better selves as they might be; that is one way of putting it; there may be no satisfactory way.

There is that moment in *King Lear*, when the broken Lear acknowledges his wrong to Cordelia, the sincere daughter he disinherited and disowned:[13]

> If you have poison for me, I will drink it.
> I know you do not love me; for your sisters
> Have, as I do remember, done me wrong:
> You have some cause, they have not.

We expect, I suppose, something like:

> Alas, dear father,
> I do forgive you freely.

What Cordelia says is: 'No cause, no cause.' On one level, a soothing murmur; on another, a forgiveness so total that it obliterates all the sin.

Pearl is linguistically so difficult that it can be a trial to an undergraduate; but one of the supreme moments in my own undergraduate reading came as I battled on through this exacting, perplexing, beautiful poem:[14]

> The court of þe kyndom of God alyue
> Hatȝ a property in hytself beyng:
> Alle þat may þerinne aryue
> Of alle þe reme is quen oþer kyng,
> And neuer oþer ȝet schal depryue,
> Bot vchon fayn of oþerez hafyng,
> And wolde her corouneȝ wern worþe þo fyue,
> If possyble were her mending.

(The court of the kingdom of the living God has a property of its own: everyone who may arrive in it is queen or king of all the kingdom, and yet never will deprive another, but each one is glad of the other's having, and would like the other's crowns to be five times as precious, if any improvement were possible. – But a prose crib loses a greal deal.)

We compete, we want status, nurse grievances, compare touchily, grudge other people their luck; and here, in the midst of apocalyptic images few people today take literally, shines an image still claiming our reverence: a kingdom pure of all envy and competitiveness, in which happy people rejoice in the happiness of others.

There are many other passages in Shakespeare; there is Chaucer's great moment when Troilus looks down from the seventh sphere and laughs that anyone should mourn for him;[15] there are certain moments in *The Prelude*, or *Prometheus Unbound*, or *Rabbi ben Ezra*, or Traherne's *Centuries of Meditations*. There may be moments less expected; for me one example is the letter of Marie Goesler, declining the proposal of the Duke of Omnium, in Trollope's *Phineas Finn*; here, in a context of worldly and calculating people, the wisdom of a difficult decision and the exquisite tact of the letter seem to me to open a window on our human possibilities.[16]

Reading about the good will not make us good; but it may throw a little light on the path.

2

FLYING SOLO

> And he reflected, dazzled by the unforeseen chances of existence: 'Yesterday I was at school – and today I see this!'
>
> Arnold Bennett, *Clayhanger*[1]

The transition from school to university, being a glorious liberation, provides opportunities for some inglorious new mistakes. Our essential adulthood is acknowledged; we are released from sometimes irksome restrictions, duties and pressures of a collective organized for children. Freedom to specialize fully in a favourite subject is intellectual luxury. The university offers, too, a variegated bazaar glittering with choices: studies, lifestyles, habits, dress, ideologies, cultural experiences, pastimes, friendships.

Most students have, at some time in the course, problems that cost more pain and anxiety than mere study difficulties. Common among these are: budgeting on what is usually a small income; accommodation difficulties; coming to terms with love and sex; religious, ideological or philosophical difficulties; career problems; family tensions; and even (odd though it may seem) loneliness. However, such problems are not unique to students, and we are scarcely men or women till we have agonized through most of them. A student ought to be able at least to *study* cheerfully – often joyfully – and fairly effectively.

He may, however, find some adjustments difficult. The subject proves to be much bigger than it once seemed. An hour in a university library may be an alarming eye-opener as to how much knowledge exists. Lecturers may take for granted background knowledge that a student lacks.

Exciting opportunities appear: lecturers know aspects of the subject with depths of specialization impossible for school teachers; discussions with fellow-students who have read books you were not told to read, been influenced by teachers you never met, and of course have minds of their own, may force much rethinking; exercises make new demands on your mind, and receive new kinds of comment and correction. Even a good bookshop is for many students a new experience. The mass of reading required is wide-ranging, enormous, thrilling, daunting.

All this broadens our experience wonderfully. So would an up-river exploration of some fertile, buzzing, whispering jungle; but a triumphant return with journal and specimens is preferable to an end with bones sticking up from a mangrove swamp. Explorers should look where they are going.

The perils of the literature course are relatively mild. Professors seldom bite. Minds are designed to expand. Even the risk that, under the influence of books, we may seriously strive for the integrity of George Eliot, the wide compassion of Shakespeare or the self-discipline of Milton, is not alarmingly great. If the student feels he has been parachuted into alien territory and is on his own, he can console himself: the alien territory is not hostile.

However, he is, and must be, largely on his own.

School discipline has gone. A lecturer will recommend good books, but will not enquire later which you have read. Few lectures or classes are compulsory; you are unlikely to be asked why you were away. University staff rarely waste much time chivvying students for overdue work. You will be given some information; you will not have the correct verdict on an author dictated. There is no final correct verdict. The task of the university is to equip you to make independent judgments, at least defensible, on a basis of accurate data and coherent thinking. This is the dignity of freedom; but we have to cope with our privileges.

We may perhaps start by scrapping the vulgar use of *academic* as a synonym for 'impractical', with the noun denoting some bloodless, boring pedant. The truly academic academic is among those who manifest how the human being is unique among the animals: he has intellectual curiosity; he has a concept of intellectual honesty; he wants to gain some truth and give out some truth. He can still have and give love, friendship and fun – to all of which lively, trained intelligence can contribute much.

We must go on to realize that much of true academic work is inevitably done in solitude and silence. The youngest student must read, alone, for hours; there is no other way to have an ample experience of literature. Discussion should broaden our minds, counteracting our crudest subjectivities; but first we must acquire material to discuss.

There are two real exceptions to the general programme of solitary reading. Friends reading poetry aloud in a group may realize skills of rhythm, rhyme, onomatopoeia and other sound effects more vividly; an informal play-reading may throw useful light on the specifically dramatic skills of a play.

To read enough, and carefully enough, at the new level of intellectual activity, demands not only a critical attitude to the literature, but quite a self-critical attitude in the reader; the discipline of solitary study depends on the student alone.

Here is a student who has been 'reading' from 9.30 a.m. to 12.30 p.m. – three hours of solid 'study'.

Now for the deductions:

Making coffee	10 minutes
Chatting in kitchen	10 minutes
Drinking coffee	10 minutes

Walter-Mittyish reverie triggered by words carrying romantic private associations	20 minutes
Finding lens cloth (used as a bookmark)	10 minutes
Polishing spectacles	2 minutes
Looking up (laudably) *tardigrade* in the dictionary and succumbing to the fascination of other words	15 minutes
Necessary trip	7 minutes
Fidgettings (assorted)	10 minutes
Little blank lapses of attention, allowing on average 0.25 min. per lapse	12 minutes (a low estimate)
Thoughts of lunch, triggered by oniony smell from kitchen	5 minutes
	1 hour 51 minutes

Leaving, for genuine attention to studies:

NET TOTAL 1 hour 9 minutes

This would not do for a sentry. But is it not how most of us sometimes 'work'?

An honest check, early in the first term, not only of how many hours in the day you have spent meaning to study, but even of how many minutes you have spent in each hour really studying, may be a shock causing rueful laughter, shame and improved self-knowledge. After all, some people are, meanwhile, driving trains, performing surgical operations or working at conveyor belts.

The first requirement for a student's lone reading is that it shall be done; and it is disconcertingly difficult for most of us to stick at a job with no one looking. One of the advantages of reading in a library is that the rule of silence and the presence of other readers make it harder to mess about pretending to work. Many students also find that they can concentrate better if they sit at a desk, on a fairly hard chair, reasonably upright, rather than curl in an armchair; but we can all help ourselves by observing what conditions suit us best. Some of us, for instance, work better in the morning, some in the evening. We cannot get through a university programme, or hold down a job, by working only at the times and in the way we like best; but if we can use our own idiosyncrasies to our advantage, instead of falling victim to them, we may be appreciably happier and more efficient.

University assignments are not usually chopped into bite-size pieces as most school work is: some piece of study, say of *Paradise Lost*, may extend over a whole term; most pieces are spread over a week or more; the student, having to manage a number of specialized studies running parallel, with no prescription as narrow as 'Tuesday's homework', but with exercises to hand in on time and a mass of reading to do in addition, may feel bewildered, even burdened.

Some general principles of priority help. Specific tasks to be produced on specified dates claim the most urgent attention. Some find it best to dispose first of the small, precisely delimited task; if you have to translate eighty lines of *Beowulf* for the next class, or prepare to discuss six cruces in *Hamlet*, or make some notes on the diction of *Euphues*, finish that and put it aside till you revise the work just before the class. The thought of it undone will not nag you, and the task is so clearly defined that you know when it is completed. (Not that any academic work is ever *finished* as is the pruning or darning; we could always do it a shade better, find one more fact; but we do know when we have done our best in a limited time.) From the small task the student can proceed to the larger specified piece of work, such as writing a weekly essay. Other students find that they work best by doing first the essay, in which there is a kind of creative element, so that it requires a fresh mind; and that something such as a translation can be done when their energies are past the peak. The great thing is to observe ourselves and find out how we really work best; and not to leave too late anything with a deadline; it is always more comfortable to finish something ahead of time and have it out of the way, than to put it off and worry about it.

Let me now add to such scientific axioms as Parkinson's Law, Murphy's Law, and the Peter Principle another, based like them on years of observation, notably of myself:

> BOULTON'S LAW: *The minimum time required for completing a given volume of academic work is invariably greater than the time first supposed by the student to be the maximum.*

Joking apart – to realize this early in the course will save a deal of embarrassment, sloppy work, stress and disappointment.

So priority goes to prescribed tasks first, for peace of mind and an unblemished record; then comes the more flexible reading programme. This is, in relation to human capacities, infinite. We could always read the masterpiece once more, or read one more critical book or article. We have to do some selecting. On this, tutors and lecturers will give some help, though often by implication more than by prescription; but we have to settle for ourselves what reading is to be done just now.

At least four levels of reading are required in university work; before we begin a particular stint of reading, it is as well to consider on what level we are doing it. This may even dictate our physical level; I could do type-4 reading reclining on my bed, but would always sit at a desk to do type-1.

1 CLOSE ANALYTICAL READING

This is required for all 'set books' prescribed for 'special study' in the fullest sense; for examining enough of an author's work to be able to describe his style; for relating a critical theory or manifesto to a book written on those principles or from which the principles are deduced; for relating a literary work to the history of language, and so on. It involves close concentration, examining the function of each word, checking meanings, pursuing allusions, considering ambiguities, and the like.

Few people can read like this for long at a time; it is exacting, tiring, interesting and highly enlightening. It shows us the vast difference between ordinary casual reading and the possible total experience of a work. We need to be fresh and alert to do it properly.

2 DIFFICULT READING FOR THE CONTENT

Most of us find abstractions quite hard to grasp. Dryden's *Essay of Dramatic Poesy*, Johnson's preface to his edition of Shakespeare, the preface to *Lyrical Ballads*, Coleridge's *Biographia literaria*, Eliot's *The Sacred Wood*, demand that we should check constantly that we understand a generalization, are following a line of argument. Some critical works are more difficult than these, because, though the content is worth attention, the style is cumbrous or jargon-filled.

Most of us, engaged in such reading, have to do it slowly and keep referring back to make sure we have really grasped the ideas. Taking notes, tabulating, paraphrasing and discussing may all help. We may need the help of commentators too.

We need to be fresh, calm, and ready to reason with words, not just drift along their flow.

3 READING FOR A BROAD IMPRESSION OF A BOOK

Only a mature Spenser specialist will have a really thorough knowledge of *The Faerie Queene*; we can scarcely hope to know all the plays of Shakespeare as well as we ought to know his greatest tragedies and comedies; we may have a valuable experience of *Middlemarch* or *Ulysses* or *Don Juan* without carrying away all the details or even all the plot. To read the *Oxford Book of Sixteenth Century Verse* gives us a general view of the poetry of that century, with some idea of the poets, popular verse forms, favourite ideas and fashionable images; but we may safely assume that we shall not in Finals be asked to differentiate between William Hunnis, Anthony Munday and Henry Chettle.

Such reading may be rather more relaxed than types 1 or 2; it is more immediately enjoyable, though those who cannot also enjoy responding to an intellectual challenge can hardly be good students. Reading for a general impression is something we can manage when relatively tired; it may even refresh us.

4 READING TO IMPROVE OUR GENERAL BACKGROUND

There is no limit to this. We never know enough about the total environments that shaped the writers we are studying.

This background reading is not very demanding except in sheer quantity and in the unfamiliar concepts that may be involved; but it is sensible to keep paper and ballpoint to hand and make methodical notes of whatever seems so illuminating that we specially wish to remember it.

A student, thrown in at the deep end, must somehow swim on his own; but nobody wants him to drown. Most students could have more help than they get.

Syllabuses, examination regulations, even question papers

of past years, are not under the Official Secrets Act; a look at these in a library may give useful warnings and reassurances. It is perverse to cut too many lectures or classes and then to grumble that lecturers do not help students – that is, do not help people who keep well out of reach. Library staff are usually helpful people; it is not kind to pester them about information that is under your nose all the time; first point that nose at a shelf or catalogue and *look*; but a student who really does not know how to find something is likely to get patient and friendly advice for the asking.

The university teacher who does not, normally, respond to a civil question with a civil answer that contributes something to unperplexing the student is, in my experience, rare. Some considerateness in choosing place and time is appreciated; someone about to give a lecture feels strained, and after giving one feels drained; but if the question is a complicated one, it is always possible to ask the lecturer for a convenient appointment. Though students should show some respect for the value of a lecturer's time, students should ask questions. The great scholar's endowments do not include telepathic perception of students' difficulties.

And, of course, the person of whom each student should ask several questions daily is the person in the mirror: what does this mean? do I understand? have I grasped this? does this mean what I think? is this sentence I have written clear? am I paying genuine attention? have I been a lazy toad today?

3

IS THE SYLLABUS SILLY?

> Above all, you will be sure to undertake at least a few things which you do not like; since the fact that this or that is distasteful to you shows the desirability of strengthening that particular part of your mentality to which the particular subject does not appeal.
>
> James Hulme Canfield, *The College Student and his Problems*[1]

> I want her to have a keen palate, inquisitive but never tyrannical. I want her to be able to eat at least one taste of anything in the world, from Beluga caviar to porcupine grilled with locusts, with social impunity and a modicum of inquisitive gusto.
>
> M. F. K. Fisher, *The Art of Eating*[2]

Preferences in literature, like preferences in food, vary, thank goodness. Options in syllabuses take some account of this, but some fixed cores are necessary; universities must not turn out people, supposedly having specialist knowledge of English Literature, whose knowledge is so gappy and perspectiveless that they are blind guides. Fair testing is impossible without some delimitation of what will be tested. Browsing at will is a leisure enjoyment for the rest of life; a degree course should resemble hard athletic training rather than strolls after Sunday lunches.

Certain symptoms should warn a prospective student off a

course in English – or any foreign language course of massive literary content.

Since the novel was not an important genre in English till at least the mid-eighteenth-century – that is, after over a thousand years of recorded English literature – a love of reading that is only a love of novels, even including difficult novels, will be inadequate.

A relish, however perceptive, for recent literature only, with too strong a sense that the past is past, and done with, and dull, is about as crippling. It does not much matter whether, at eighteen, our relish for the past is escapism (excusable in all), or a liking for antiques and quaintness (fun), or some beginning of a real awed sense of history and human unity (difficult and stretching towards infinity); but if we can find little interest in anything much before our own birth dates, the course will be mostly yawns and headaches.

A slow reader should think carefully before embarking on the course; the sheer bulk of reading may prove too burdensome. Reading speeds improve with practice and strong motivation; but a real tortoise-pace may be too great a handicap.

We all have blind spots somewhere; I still cannot make myself fully enjoy Thackeray; but to be, like Old Gobbo,. more than sand-blind may be as fatal a handicap. One student may find Spenser so leisurely, so alien in his ideas, that he can find nothing to savour in Spenser; another may fail to respond to Pope . . . or Donne . . . or Byron . . . or Browning . . . or Tennyson. They are all wrong in this detail, as I am wrong about Thackeray; but we cannot all respond to everything all the time. However, the reader who finds Spenser *and* Pope *and* Donne *and* Byron *and* Browning *and* Tennyson unrewarding would be wiser not to start a course on literature; his range of response is too narrow.

The time to look most critically at syllabuses is before applying for a university place. The English Association regularly publishes a *Guide to English Courses in the Universities* which covers all the possibilities in Great Britain and Ireland. The nature of the syllabus may be a factor in the choice of university. Afterwards, habitual grumbling is usually energy-wasting excuse-making.

Students have three common grouses.

First, the syllabus is too big; that is, there is too much to read.

There is always some truth in this, in the sense that work is hard. University life is so crammed with interesting possibilities that study may seem a secondary activity. As soon as we realize that the course is meant to be – and in fairness to everyone outside ought to be – about as demanding on our time as a full-time job, the syllabus looks more reasonable. And this is, or should be, the work we most wanted to do – by no means everyone's lot.

Since most undergraduates do eventually graduate, it must follow that examinations and assessments apply the requirements of the syllabus in ways, and with standards, reasonable for the average undergraduate. He is not expected to have read, still less memorized, every page that could be claimed as relevant. Past examination papers give a good idea of the kind of grasp expected. It probably includes a broad general knowledge of many fields, with detailed analytical understanding of some representative samples. Undergraduates may even be allowed to get away with a touch of bluff now and then; but, though a tactically clever question in class, a well-managed sequence of argument or an attractive style in a paper, may sometimes eke out scanty knowledge, there must be something to eke!

Second, almost every student finds something uncongenial in a syllabus.

Since tastes differ, this is inevitable; a few options can be made available, but a syllabus to suit all students in all respects is impossible. This, however, is not the whole story.

Those who really love literature, with a depth, generosity and fidelity as yet outside any undergraduate's experience, are anxious that a course in it should not be, as shallow detractors still sometimes assert, a soft option; any training worth having includes some rigours.

Syllabuses are devised by committees, and committee-work often brings out the worst in people: mean little power-struggles, contrarinesses, jealousies, horse-trading, miniature politics; sometimes a necessary compromise may be a far from ideal solution; but even bickering professors normally still care about their subjects and their students; and anyone partly responsible for a syllabus knows at least far more than any beginner about the scope of the subject.

A syllabus is intended to cover some functions of a course that may not occur to the beginner. Serious study of literature must include some kind of historical approach: how did this genre begin and develop? against what was this group reacting? were there forerunners of this major innovator? who influenced whom? how general was this trend? what could this author have read? A sound historical approach may necessitate some study of, for instance, *A myrroure for magistrates*, or *The Spanish Tragedie*, or *The Seasons*, or *The Mysteries of Udolpho*.

There is little point in taking the course if it does not considerably enlarge our sense of the vastness of the subject. I found my first degree course a revelation comparable to the view from an aircraft after some walks and coach trips.

> Hills peep o'er hills, and Alps on Alps arise![3]

Often a student needs to say to himself what a reasonable, not a bullying, parent says to a face-pulling child over a new dish: 'How do you know you don't like it until you have tasted it?'

Two experiences are wonderfully educational: to meet something we love, on a new level of attention, and find there is far more in it than we ever dreamed; to be forced to encounter something we believe is unalluring, and to find, through real attention, that it is full of exciting merits. I loved *The Rime of the Ancient Mariner* when I was twelve, could still have a new experience of it through a superb lecture heard when I was fifty-two.[4] I tried Henry James when too immature, might never have tried again if I had not needed to, and then found his awe-inspiring craft, subtlety and perception.

Any fool can be dismissive; nothing is easier, or more cheaply shields our own conceited self-satisfaction, than to reject everything we at first find perplexing, uncongenial or dull. The history of criticism is pocked with the bad shots of critics (most of them far more experienced in literature than students) who chose the smug and easy way, insulting, among others: Keats, Shelley and Wordsworth; Browning and Tennyson; the Brontës and Dickens; Hardy and D. H. Lawrence; Auden, T. S. Eliot, Robert Graves and Yeats. Blake was virtually ignored in his own day and not reviewed. It takes a wise man to look at something new hard enough to see it. A syllabus that yanks my head round and makes me look in a fresh direction gives me a chance to see strange sights.

A quite sensitive student might argue that, with an abundance in English of really major works inviting attention, many syllabuses give too much room to the second-rate, say Sidney beside Spenser, Gower beside Chaucer, even Jonson beside Shakespeare or Smollett beside Fielding. However, as most of us would be about twenty-second-rate, we need not turn up our noses too disdainfully at the 'second-rate'. Virtually no literary work that survives, even if chiefly for specialists, is without some intrinsic interest and value; and it is often the relatively 'second-rate' that we should never otherwise sample for ourselves.

Moreover, when we have read the plays of, say, Jonson, Webster and Ford, we shall not only have found much eminently worth reading, but understand far better why Shakespeare reigns supreme. Too much comparing, as if we set ourselves up as examiners to put our betters into class lists, can become an exercise not only barren, but stupidly presumptuous; but we do appreciate the geniuses of any epoch better against a background of lesser talents, competent craftsmen and fashionable imitators. Lydgate was far from worthless; read before judging; but if you are in any doubt about the calibre of Chaucer, the miracle of his achievements in his epoch, read Lydgate.

Then there are works, perhaps not immediately exciting to the twentieth-century reader, that are key works in the history of English literature – in the development of literary techniques: *The Shepheardes Calender*, *Absalom and Achitophel* and *Waverley* are examples.

A third objection to a syllabus may be that it excludes something we find congenial and believe to be good. This may be the grievance of the better student, wanting more, not less; but it is also the least disquieting grievance. The scope of English Literature is so huge that the most devoted specialist in it knows he will die not having read everything he would wish to have read. A syllabus may make a student read something he otherwise would not read; it can never for long stop him studying anything he wants to read!

I did my share of grousing, as an undergraduate, about Old English in the syllabus; for many students the rewards seem small in proportion to the efforts; yet a specialist course in English that ignores the first three hundred years of literary history and the foundations of the language is highly questionable. It was while I was working on Old English and a whole syllabus that in my youth was still confined to literature before 1830, that I was also rapturously discovering the poetry of the 1930s. We are *compelled* to read something only by the syllabus of the subject we have ourselves chosen; it is one of our enormous pieces of good fortune in Britain that there is no work of English or world literature we are *forbidden* to read.

The student who resents an uncongenial part of the syllabus may well wear a kaftan, enjoy Chinese food, long to see Katmandu. Life is made more interesting by the widespread liking for the 'ethnic', which forty years ago we would have called the 'exotic'.

All such curiosity brings new breadth and savour to life; but Old English studies offer a startlingly exotic experience of a culture far more distant and alien than modern Japan, yet part of our own roots; medieval, Renaissance, Augustan, Romantic, even Victorian literature invite us to experience much that is remote and different; while, as during foreign travel, we find ourselves responding to the recognizable, touching, mysterious humanness and likeness to ourselves of these alien personalities. Our studies become happier as we learn to appreciate the demands of the syllabus as invitations to exploration, new experience, adventure.

Exploration needs to be both wide-ranging and at times minutely particular. Serious study includes experiences resembling those of personal love: true and willing attention; delight in the object for being itself and other than ourselves; the respectful wish to treat it well. Study resembles affection also in this: until we have loved someone with intense awareness, concentrated passion, real curiosity, sympathy, caring and commitment, we know little of the heights and depths of feeling, our own strengths and shortcomings or the awe-inspiring mysteriousness of the human heart. Yet our experience is also too limited until we have also felt interest in a good many other people, some kind of lesser love for many relatives, friends, acquaintances, colleagues.

One function of the 'set text' is to ensure that what the student selects from a voluminous, uneven corpus, of which he may be unable to read the lot (Chaucer, Spenser, Milton, Jonson, Dryden, Byron, Wordsworth, Dickens, Browning, perhaps even Auden) shall include something representative of the author's best work, say *Othello* or *Twelfth Night* in preference to *The Two Gentlemen of Verona*; *The Alchemist* rather than *The New Inn*. Another function is to ensure that the student shall have a truly full experience of some literary work.

A friend tells me that students sometimes ask her – not ironically – 'Do we have to read all the set books?' She manages a kind but firm affirmative; many lecturers would be tempted to a reply such as 'cannot with propriety be read aloud in a family'.[5] The set books are a minimum collection for minutely particular study, involving thorough knowledge of the text itself, close analysis, annotations, reading of relevant critical works.

A set book needs to be *set* not only in the syllabus, but in its context. This is true most of all when it is a portion of a much larger work. A close study of *The Prologue* and *The Pardoner's Tale* includes a quick general reading of *The Canterbury Tales*, not necessarily checking every unfamiliar word in the glossary or closely following the arguments through *The Tale of Melibeus*, but seeing how Chaucer fits tales to characters, uses different genres of tale, creates a temporary, random, heterogeneous community; we need to see the set texts as specimens of greater riches. To study Books IX–XI of *The Prelude* we need some knowledge of the French Revolution, of certain classical and other allusions, some bits of geographical information; but most of all some knowledge of the rest of *The Prelude*, to understand how the psychological story develops and with what techniques it is expounded. Similarly, a set portion of *The Faerie Queene*, *Paradise Lost*, *Don Juan*, needs to be related to the total work.

And, though *The Two Gentlemen of Verona* would be a freakish choice for a set book, any student of literature should during the course have read every Shakespeare play at least once; that corpus is the supreme volume in English literature;

it also throws a flood of light on many varieties of dramatic and poetic technique.

Syllabuses, and lecturers' advice on interpreting them, will indicate a minimum programme of wide-ranging, more impressionistic reading for the specific course; there is no maximum. We can never have read too many different works of literature; each one adds something to the total perspective; and we can never have studied a representative text too closely.

4

DO WE MURDER TO DISSECT?

> Men are admitted into Heaven not because they have curbed & govern'd their Passions or have No Passions, but because they have Cultivated their Understandings. The Treasures of Heaven are not Negations of Passion, but Realities of Intellect, from which all the Passions Emanate Uncurbed in their Eternal Glory.
>
> William Blake[1]

> Sweet is the love which Nature brings;
> Our meddling intellect
> Mis-shapes the beauteous forms of things:–
> We murder to dissect.[2]

These lines of Wordsworth have often been quoted against analytical study of works of art. Wordsworth was not contrasting the commentary with the masterpiece, but a 'spontaneous wisdom' absorbed through harmony with Nature, with second-hand experience from 'sages'; and this was only one mood; Wordsworth himself wrote books he hoped might pass on some wisdom, and paid magnificent tribute to literature in Book V of *The Prelude*. However, when people eagerly snatch a quotation, even misapplying it, it must seem to them to express happily something they strongly feel.

Inexperienced students often say, 'It spoils things to analyse them!' Pope mocked[3]

> Thy mighty Scholiast, whose unweary'd pains
> Made Horace dull, and humbled Milton's strains . . .

and Samuel Johnson, himself an outstanding scholar and critic of his epoch, wrote scathingly of arrogant annotators[4] and admitted that 'Notes are often necessary, but they are necessary evils. . . . The mind is refrigerated by interruption . . . the reader . . . at last throws away the book, which he has too diligently studied.'[5] The scholar often has a bad press: Holofernes, Jonathan Oldbuck, Casaubon; it is generous Browning who in *The Grammarian's Funeral* shows us some glory behind apparent futility.

Yes, there can be a disproportionate pedantry that cannot see the wood for the trees; there are a few emotionally anaemic people who burrow into minutiae of scholarship away from the great challenging works rather than into them; there can be bad scholarship just as there can be bad dentistry or accounting: careless, dishonest, contentious, needlessly tiresome to read. Yes, there is something of a critical 'industry'; academics, like the rest of us, appreciate money and status, academic employers often want research degrees and published articles; so sometimes people produce work in which they are not passionately interested and which is of little value. But these are imperfections sometimes found in relation to scholarship, not scholarship. (It is perhaps significant that George Eliot's Casaubon is eventually shown to be ignorant of important work in his chosen field – his desiccated life is not even one of true scholarship.)

Students are sometimes turned against analytical study by poor and premature forms of it at school. Schools never have enough money for books, so books have to be 'made to last' until all the flavour has been chewed out of them. Our experience of a good book during a lifetime is almost limitless; every time I read a Shakespeare play or *The Prelude* or *Bleak House* I can find some new excellence in it; but our experience over a short time in childhood is bound to be limited; our own experience and perceptions have not developed. The young adult beginning to specialize should be ready for real intensive study; but too much of being *kept busy* over a book may have impaired his appetite.

When we have admitted that there are possible falsities and irritations, we should not lose sight of the real purpose of intensive study: it is like the intense attention of love to a beloved person; we hope to attain not boredom, but a deeper intimacy.

We may fear bringing cool intellect to bear upon the work, thinking that this will block our spontaneous emotional response; that we cannot both think and thrill.

This is a needless anxiety based on a false dichotomy. Real thinking has its own thrills. Sometimes they are even comically physiological. Students, and experienced scholars, making research discoveries, encountering new ideas, organizing coherent arguments or receiving them, suddenly seeing how a number of data fit together in a pattern, quite often wriggle, squeal, exclaim, gasp, dance, even have mild orgasm, at the intellectual climax. I have often shed tears from sheer mental excitement in a lecture or seminar. Curiosity is a noble passion in the scholar, as it is a base one in the censorious gossip; but, made glorious by good motives and purposeful disciplines, it is still a passion.

It is simply not true to general experience that understanding spoils appreciation.

A friend will not find a pork dish I sometimes cook less palatable if she can identify the fennel, garlic and juniper berries that give it a special flavour. I can see a football match only as some men kicking a ball and one man blowing a whistle, but if I knew the rules and understood the tactics I should probably admire plucky and clever play. Seeing the sights in a foreign town is far more rewarding if we know some rudiments of the relevant history and folklore. Surely (unless there has been gross deception) we usually love friends and sweethearts more, not less, as we come to know them better? and we often even feel less bitter towards our enemies as we come to understand them more. All too many of us can bear witness to the incalculable amount of avoidable suffering caused by the superstition that information and forethought concerning sexual activities will ruin a wonderful mystery or take away happy spontaneity.

Our first, spontaneous, eagerly delighted, or even antagonistic, reaction to a literary work is genuinely important. It must be where we start from. Yet we all correct first reactions, even with no specialized studies at all; a novel that means nothing to a fourteen-year-old may greatly please him at forty, simply because his experience of life has extended to include the emotions portrayed; and he may now dismiss, as too foolishly impossible, a tale that he loved at fourteen.

Closer study may eliminate actual misunderstanding – of words, of sentences, even of the whole book. We may at first have missed allegory, irony, important allusions; we may need a biographical, historical or ideological context; not only have words changed their meaning, but human societies have changed and so has the collection of ideas usually taken for granted. However we are to interpret *The Taming of the Shrew*, we must not see it as a counterblast to women's-lib concepts wholly outside Shakespeare's experience; *The Tale of a Tub* is virtually unintelligible without some idea of the theological controversies of Swift's time and its many preliminaries and digressions, often brilliant, are not funny until we have some notion of the things they satirize.

We may gain a great deal more from a book by placing it in its context. While we can to some extent appreciate the literary jest of *Northanger Abbey* if we compare Catherine's addiction to gothic fiction with a modern addiction to horror films, we shall have a much more detailed enjoyment of Jane Austen's fun after a look at a novel by Mrs Radcliffe. Much in

In Memoriam appeals at once to any troubled, questioning mind or bereaved heart; but a deeper understanding comes with some knowledge of Tennyson's probable reading background and the climate of opinion in which he was writing.

Closer study may reveal – or a helpful expert may point out – details of craftsmanship that enhance our enjoyment when we observe them. A lecturer's reading aloud may show up apt sound effects in a poem; a closer look at a metaphor may reveal a subtlety of appropriateness not instantly obvious; scanning a blank-verse dramatic speech may show how skilfully the rhythm helps to represent the details of emotion; re-reading a novel, we may notice many small touches of action and speech that go to building up the impression of a personality; we may catch the significance of a bit of parody, imitation or quotation; we may realize the ambiguity of a word or phrase and from this understand new, rich complexities; looking up a name in a classical or biographical or biblical reference book sometimes does no less than show us the whole point of a verse or sentence; and so on. Such insights cannot possibly detract from our pleasure; they can only increase it.

There is one approach that really can temporarily spoil a literary work for us; one to which students, with assignments to do and examinations to pass, are sometimes tempted: to read some critical commentary before – or even instead of – the work itself. As that most endearing of pioneer critics, George Puttenham, would have it, 'we call it in English prouerbe, the cart before the horse, the Greeks call it *Histeron proteron*, we name it the Preposterous . . .'.[6]

Our own experience, however sloppily subjective, however limited, should always come first. Writers do not write with the intention of producing examination material; they write to give pleasure and communicate their views. Johnson, as so often, gives us eminently commonsensible advice:[7]

> Let him, that is yet unacquainted with the powers of *Shakespeare*, and who desires to feel the highest pleasure that the drama can give, read every play from the first scene to the last, with utter negligence of all his commentators. When his fancy is once on the wing, let it not stoop at correction or explanation. When his attention is strongly engaged, let it disdain alike to turn aside to the name of *Theobald* and of *Pope*. Let him read on through brightness and obscurity, through integrity and corruption; let him preserve his comprehension of the dialogue and his interest in the fable. And when the pleasures of novelty have ceased, let him attempt exactness, and read the commentators.

This is the right sequence of experiences. However, the student, in responding to Johnson's sympathetic sanity, should respond to *all* of it; it is the task of a specialist student, in due course, to 'attempt exactness'.

A set book must usually be studied in an annotated edition. An example is Browning's *Men and Women* in Paul Turner's edition (1972), in which a brief introduction and biographical chronology place the book generally; the text used is identified; and seventy-five pages of notes give the three kinds of help such notes normally supply.

One form of help is annotations on allusions. The student is expected to understand every detail of the small number of texts prescribed for detailed study; but he can hardly have acquired the wide culture of the mature Browning; he ought to know who Titian or Lazarus was, but it is entirely excusable – until *Men and Women* has been studied as a set book – that he should not identify Camaldolese, Karshish, Guizot, Escobar or Terpander, or understand Andrea del Sarto's 'cue-owls' or the rare sense of *male* in the 'male-sapphires' of *Saul*. Such annotations – not always written as gracefully as Turner's – may seem to demand a lot of time; true, they do; a close study of a complex work is bound to make rigorous demands on us; but if we seek the total experience the author intended to give, we need the information; and how long would it take each student to dig it all out for himself?

Secondly, notes can help us to follow something that is difficult, not because of rare references, but because of an elliptical style, an unusual thought-process, and so on. Turner has a long helpful note on the mysterious *Childe Roland to the Dark Tower Came* but perhaps it is most helpful in the comforting admission that, in spite of the efforts of numerous scholars, no totally adequate explanation exists. Or there is a paraphrase of the lines from *The Statue and the Bust*:[8]

> Where a button goes, 'twere an epigram
> To offer the stamp of the very Guelph.

which anyone may be forgiven for finding difficult.

Third, annotations may enlighten us as to the biographical or historical background of a work: for instance, if we read *Cleon* without Turner's detailed specialist knowledge, we undoubtedly have an interesting experience, encountering an imaginative impersonation of a pagan intellectual troubled by the ultimate futility of 'progress' and 'immortal' achievement with no personal future life, with references to Christianity that have several layers of irony. If the first experience were not interesting (to the careful reader), the poem could hardly be regarded as successful. The notes, however, enrich our experience in two ways: by placing *Cleon* in relation to 'the religious difficulties of Victorian intellectuals'[9] and by explaining some details of Cleon's cultural background.

Annotations may in fact lead us to many intellectual thrills, sometimes even those 'Gleams like the flashing of a shield'[10] that art and learning, as well as natural beauty, can give us: in our alternations between depth and breadth, we find that depth and breadth are one.

An intensive study of Thomas Browne's *Hydriotaphia*, as a conveniently short example of seventeenth-century ornate prose, will require us to pick up numerous biblical references; aspects of ancient history and ancient historians, and later historians now little known; Greek and Latin; Roman literature and philosophy; early British history; exploration; non-Christian religions, and Christian traditions no longer familiar; Dante, Cardinal Farnese, the Phoenix. We should be able to rouse ourselves to some mild antiquarian curiosity and enjoy gratifying it; but the real dividends are bigger. In thoroughly studying *Hydriotaphia*, we have, as we soon find, equipped ourselves much better for reading Shakespeare, Milton, Donne, Byron – right up to Auden and Peter Levi. The Roman historian Diodorus Siculus, whom Browne quoted in 1658,[11] is not widely read today; but Seamus Heaney quoted him in 1975.[12]

We come to new awarenesses of a great accumulation of culture available as a common stock; in our study we find an awe-inspiring, often touching unity of human experience, transcending individual circumstances and even epochs; and an even more astonishing and touching diversity; the same reference has stirred many imaginations, but the exact response, the imaginative interpretation, has always been different. Scholarship moves by the slow pains of striving for accuracy; but it can end as it began, in delighted wonder.

Any real love, not only of an art, but of a cause, a sweetheart, spouse or friend, of food, wine, flowers, jewels, vintage cars, is manifested largely by attention, that remarkably central virtue, so dull when in school it means not looking out of the window, so vital when it means really listening to someone we love. How often does love fail, not from ill-will, but because we have not really listened to what someone said, not truly grasped the need or the difficulty, concerned ourselves more with our own muddled instant subjective reaction than with really reaching out in intelligent sympathy to the other person? *Mea culpa*; my fault indeed, but a fault almost universal.

Close examination of a literary work should be essentially a loving activity; an attention reaching out towards a fully perceptive intelligent sympathy. Since we are unlikely to have at the same moment the spontaneous response of an almost passive surrender to the work, and the very active intellectual exercise of analysis of the work (and of our perhaps imperfect reactions to it), we may say that sometimes we anaesthetize to investigate; but we do not murder to dissect. The work well explored takes on a more vivid life. Ultimately we shall wonder more delightedly at what we have come to know better.

5

RELEVANCE AND REVERENCE

> The orator yields to the inspiration of a transient occasion, and speaks to the mob before him, to those who can *hear* him; but the writer, whose more equable life is his occasion, and who would be distracted by the event and the crowd which inspire the orator, speaks to the intellect and heart of mankind, to all in any age who can *understand* him.
>
> Henry David Thoreau, *Walden*[1]

> We're none of us infallible – not even the youngest among us.
>
> William Hepworth Thompson, when he was Master of Trinity College, Cambridge[2]

For some years it was fashionable among students, especially students whom often generous and unselfish motives had led into ideologies whose spitefulness and barbarism were not immediately obvious, to dismiss many topics of study as 'irrelevant'. Fashions in wording pass; our human capacity for misunderstanding does not.

That particular term was inept; a book can no more be *relevant* than it can be *longer*; it can only be longer *than* another book, and relevant *to* some other matter. *Silas Marner* may be highly relevant to my temptation to retreat from human relations into nursing my grievances, if I have been let down and unfairly treated; it is irrelevant to my problem of what to wear at a party; it can be neither relevant nor irrelevant in thin air.

Students to whom social and political commitments seem supremely important are tempted to dismiss many kinds of literature as obsolete and worthless because they are not overtly relevant to urgent issues such as 'the liberation of mankind', 'the class struggle', 'fighting colonialism', 'racism', 'women's lib', 'alternative living', 'the environment', or whatever is the current concern.

Such a question as 'Ought I to spend time on Donne's once new techniques of metre and imagery, when today new techniques of torture are being used on innocent people held in atrocious conditions over much of the world, under capitalist, marxist, nationalist, tribal, racist, religious or personal dictatorships?' is not a silly or ignoble question. If it fills so much of your mind that an English degree seems almost wickedly luxurious, perhaps you should not be reading English, but working full-time somehow to fight this appalling evil. You could thus become more like a saint than I ever shall.

Most of us who find such a question disturbing compromise, as I do, by some additional activities such as supporting Amnesty International and sometimes writing letters of protest; if we all did even so little, it would help a lot.

Is *She Stoops to Conquer* a mere frivolity, when many politicians (or tycoons, or trade union negotiators) will stoop to almost anything to conquer? What value lies in the virtuosity and linguistic excitement of *The Shepheardes Calender* when millions of people in backward agricultural communities survive, if at all, scraping a bare subsistence? Why bother with the subtleties of *The Ambassadors* when it is desperately important that nuclear or biological warfare be prevented?

To such questions there probably cannot be an answer so total and final as to license a fully comfortable complacency.

Chapter 1 gives or implies some fragmentary answers. A few more points may be made.

The crudest is that a student who perseveres to a good degree has probably started on the way to a position where his influence and resources for doing good will be much greater than those of the drop-out, unless the drop-out is of about the calibre of St Francis of Assisi.

Then, too, political and social questions are genuinely important, but the whole community cannot spend its whole energies on them, or there will not only be nothing for *Notes and Queries*, but no bread, bricks, bathwater, baby clothes, bandages or ballot boxes. I hold a firm conviction – probably derived as much from reading good literature as from anything – that everyone should try to do something to reduce the volume of suffering in the world; but sometimes even a good cause can obsess us till it becomes a psychological disease and we do more harm than good.

Third, literature is not irrelevant to politics and society just because it says nothing in so many words about, say, 'the class struggle' or 'the Red Menace' or 'inflation'. Sometimes the insights of great literature may come nearer to the sad complexities of life and truth than the pamphlet whose purpose is to argue a case. What Shakespeare shows us about the limitations of government and the imperfections of human justice in *Measure for Measure*; about the differing narrownesses and unawarenesses of patrician and populace in *Coriolanus*; what Jonson shows of greed and gullibility in *The Alchemist*; what Wordsworth in *The Prelude* or Carlyle in *The French Revolution* shows of how crying injustices can lead to screaming revolt and how easily a justifiable revolution can be warped into a bloody tyranny, are still today as painfully relevant to current affairs as anyone could wish. Productions of Shakespeare in modern dress have often emphasized such continuing relevancies.

Yet, finally, literature does not have to justify itself by

relation to a primary task of solving or even exposing current social questions; its main business is to show us some of the more permanent mysteries and multiplicities of our human nature and social relations. Topicality wins keen temporary interest and looks like bravely 'committed', 'relevant' work, but nothing sooner becomes irrelevant; nothing dies sooner than the most specifically topical. The *Immortality Ode*, *The Prelude*, *Michael*, survive; what of Wordsworth's sonnets: *November 1813*, *Upon the Late General Fast* (1832) or *Protest against the Ballot* (1838)? The topical poems by Wordsworth that are still quoted are those notable for exceptionally impressive phrasing, such as 'Once did She hold the gorgeous East in fee' or for an element of universality, such as *To Toussaint L'Ouverture*, which now needs a footnote on its once topical subject, but is in some sense about all good men in defeat, all prisoners of conscience.

Of all English poets who may be regarded as 'major', by criteria of intelligence, craftsmanship, originality, vitality and substantial output, surely the one most nearly forgotten today except by specialists is Dryden. It is a sad shame; one can scarcely open his collected poems without finding something at least clever, probably something tougher and profounder than the merely clever; but Dryden was so up-to-the-minute in his own day that now poems as brilliant as *Absalom and Achitophel* or *The Hind and the Panther* need so much annotation to be intelligible that their magnificent energies are deflected. Again, Louis MacNeice, a poet of marvellous craftsmanship and broad, humane sensitivity, wrote in 1938 an outstanding topical poem, *Autumn Journal*[3] that in the 1940s was one of my most loved companion-poems; it remains eminently worth reading, but as time goes on more and more of the references recede; in 2038 it may have to sink under a weight of explanations like Dryden's *Annus Mirabilis*; a sad shame, again, if it must.

Wrongheaded and miserably circumscribing though

judgments of literature based on its percentage of overt approved political content are, in young individual students they have some smatch of honour in them, for they probably spring from altruistic, if often naive, ideals. (Imposed as policies by totalitarian régimes, such judgments are mere intellectual betrayal of peoples forced into blinkers.)

There are more purely egotistical kinds of choosiness and dismissiveness of which we are all probably guilty at some time or another: essentially, refusing our attention to some literary work, because it does not appeal to us immediately, undemandingly, and just as we are now.

No one appreciates everything. Jack Sprat has some right to eat no fat, though not to make a fat-free diet a general rule of decent civilized eating; but he must also accept that it is not the function of every item on the syllabus to speak instantly to the condition of Jack Sprat or Jill Sardine. After all, what we find most immediately and easily congenial may well be what teaches us least.

My mother used to fetch library books for a neighbour. The task of choosing novels was made more difficult because this poor old creature would at once push aside any book in which she spotted any explicit reference to sex. We had reason to believe she had spent much of her life in a bleakly uncarnal marriage. Yet in cutting herself off from an important aspect of life even in her reading, she was refusing the last possibility of understanding a little of how important sex might be to many people; and, generally good-hearted, she was stunting her own possibilities of charity, compassion and even relish for the variousness of life. The student who would snigger at this poor old prude might himself be reluctant to study *Paradise Lost* because he was an atheist; or *The Revolt of Islam* because Shelley called himself an atheist; or *Sonnets from the Portuguese* because he got sick of sonnets when working on the sixteenth century; or *The Heart of Midlothian* because he doesn't like dialect in novels; or *The Doctor's Dilemma* because his father is a doctor and Shaw is disrespectful to the profession; or, or, or . . . all the time limiting his own experience and shuttering some of the windows of his mind.

Naturally we do have to submit art to the test of our own experience, at least in judging its truth to life; as, for instance, I am now confident that George Eliot's portrayals of human society and motives are more closely and broadly real than those of Rider Haggard; but, when tempted to be quickly dismissive, we all ought to remember that our own experience is bound to be very limited and can never provide a conclusive test.

I once heard a youth who had probably just left his A-levels behind say, chattering before a lecture, 'Byron – oh, was he a poet?' He obviously thought he sounded clever, discriminating, and immensely grown-up.

A thuddingly dogmatic acquaintance once told me a broad mind was inevitably shallow; neither this nor its converse, that a narrow mind is deep, is true; nor indeed is a broad mind necessarily deep; it all depends on the total volume of the mind. But to be proud of having a mind too little to find room for Byron is, especially in a professing student of literature, downright perverse. In literary studies, as in human relations, we can be so eager to seem knowing, that we are unable to reach any worthwhile knowledge. We can be so anxious never to be taken in, that we can end by being shut out from every sanctuary. We do not measure a man's social adjustment, or his deeper wisdom, by the fewness of his friends.

Obviously, literary criticism must eventually include some discrimination and comparison, some power to identify intellectual dishonesties, stylistic faults, emotional falsities; but all early literary studies – and all studies of new material at any time – should begin with respect. Shaw, iconoclast though he was, made a favourite character say, 'Sneering doesn't become either the human face or the human soul.'[4] A readiness to sneer and dismiss, a suspicious, fault-finding initial approach, a patronizing attitude to the work before us, may be evidence not only of a want of common-sense humility, but of a sickly self-protectiveness: do we want to put up a barrier against the work, in case its impact, its effect, intellectual, emotional or moral, may really mark us?

It may perhaps be because children can so easily experience adults as mainly fountains of disapprobation, and overhear them as mainly grumblers, that we are tempted to think it grown-up, a mark of mature, discriminating taste, to like as few works as possible. There was a medieval bishop who was wiser when he noted three stages in increasing wisdom, '*spernere mundum, spernere sese, spernere nullum*' – 'to despise the world, to despise oneself, to despise nothing'.[5] We may have to go through a phase of spotting the 'phonies', who in my hearing have, on some lips, included Marlowe, Milton, Gibbon, Keats, Shelley, Wordsworth and Yeats; of seeing through Shakespeare and despising Dickens; but we should try to keep that phase a brief one.

I think my ideal critic would be the one who, stuck on a wet day in a seaside guest-house, with nothing to read but a volume of Patience Strong and an espionage thriller, could pick out some lines in the Strong having some kind of neatness and good sense, and see the skills as well as the sillinesses in the thriller.

C. S. Lewis happily worded what seems to me a central sound principle:[6]

> The first demand that any work of art makes upon us is surrender. Look. Listen. Receive. Get yourself out of the way. (There is no good asking first whether the work before you deserves such a surrender, for until you have surrendered you cannot possibly find out.)

Pope, long before, gave us another reminder: the fault is not invariably in the matter being read:[7]

Those oft are stratagems which error seem,
Nor is it Homer nods, but we that dream.

A wholesome exercise for the man who remains unimpressed before a cathedral is to make a rabbit hutch. From that attempt he may realize that to build not only a cathedral, but the plainest, smallest terrace house, requires many skills and much precision. Not only is Chaucer a master; Lydgate or Cowley or Cowper commanded skills considerably greater than most of us will ever attain.

Consider what an effort is demanded to produce that measly little essay in the course work – below publication standard, unoriginal, perhaps marred by mistakes of fact or even grammar, probably about as stylish as a coalshed. Try to write a minimal sonnet – say something meaningful and sequential while obeying all the rules of formal pattern. How long do you need to write three Spenserian stanzas, without beauty or vitality or the stamp of unique personality, but saying something while sticking to that pattern in every detail? What about a trifle of a short story, a mere one-act play, with a real plot and in good clear English?

And how would any of us like to write even a weekly essay or a letter home, with a quill pen that needed endless dipping and frequent reshaping?

We learn most when we approach something with eyes open, and nose not pointing up in the air, that is, away from the object, but straight at it, in the direction of full and friendly attention.

6

OWNING, BORROWING, CONSULTING

For him was lever have at his beddes heed
Twenty bokes, clad in blak or reed,
Of Aristotle and his philosophye,
Than robes riche, or fithele, or gay sautrye.
Geoffrey Chaucer, *Prologue* to *The Canterbury Tales*[1]

Books are the raw materials of the literary student's work; and his most important materials are never the *Handbook to* nor the *Survey of* nor the *Notes on*, but the piece of literature itself.

Of course we should read the critics and biographers, attend the lectures and discussions; but the most important study we can make of *The Prelude* (or *The Pardoner's Tale*, or *Pericles*, or *The Plumed Serpent*) is to read the work itself, and then read it again, and, after reading some criticism, to read the work once more, and so on. This can scarcely be over-emphasized; it is the basic principle of all literary study that has any integrity or educational value.

We should own copies of as many of the classics we are studying as possible. We can re-read the books we possess as much as we like; we can take them to classes; we can also mark them, interleave them with copious comments (especially useful for set texts), copy in information from other editions, file a few useful press-cuttings (such as reviews of new critical works) in them. They will be handy for revision as well as for verifying quotations.

Moreover, most of us probably appreciate literature more

if we actually possess many of the books we admire; as in good human relationships, respect often increases parallel with familiarity. There is some value in simply owning books, looking at them with pride, handling them, showing them to friends; even the smell of a book we have just bought (perhaps at some sacrifice) can help to give us the sense of having a stake in this world of letters; we make it less alien, more ours; we are building a little nest of our books as well as our clothes and mugs and posters.

Owning the books we should be reading most thoroughly also helps with sheer immediate practicalities of self-discipline: we are more likely to make ourselves read and re-read if we do not have to go to the library or beg the loan of a friend's copy to do it. Besides, we can read our own books over solitary meals, while drying hair, in bed, in hospital if we are unlucky, on holiday if we are lucky.

Contrary to persistent mythologies of envy, most students are comparatively poor. Obviously some priorities have to come even before books: notably obligatory fees, a place to live and something to eat; but the student of literature ought to see some books as nearer to necessities than, say, beer or lipstick. (Part of the state grant is intended specifically for book purchasing.) This is not preaching from a velvet-lined pulpit; during some of my student days I was cutting down on food and heating, and always darning, which I detest, to buy books. I do not regret it; I still have those precious volumes. There is a degree of poverty, rare among British students, that is destructive and a shame upon any society; but a few sacrifices for what really matters to us are quite dignified, testing our maturity and realism.

The student who wants to own books and is short of cash can often help himself with a bit of enterprise. He must never help himself in the jocular sense; shoplifting is unfair to everyone else; shops now usually prosecute to deter other thieves; a conviction is at the very least a humiliation, distressing also family, friends and college authorities; at worst it may mean heavy penalties, the end of a degree course and the ruin of career prospects. Modern photocopying techniques also give rise to the modern temptation for students, illegal photocopying; this will be dealt with later.

Students can help themselves in sensible, ethical ways. Money can perhaps be earned: the Students' Union may know of openings; a university bulletin, a local paper, a shopwindow card, may offer some possibility. Two things we can sometimes sell are: the grit to do a job few people want to do, such as heavy digging; some unusual skill, such as a talent for carving little wooden animals.

Money or book tokens can be won; it may seem fatuous to enter a crossword competition, but it is not at all fatuous to win yourself, innocently, the price of your complete Browning. More important, universities, colleges and some outside cultural institutions award prizes for essays related to your subject; such a win not only brings money for books, but looks well later in a *curriculum vitae*; and attempting such a task is valuable experience even for the losers.

However, a student who is giving proper attention to his studies and even a decent minimum of attention to family and friendship does not have much time for picking up extra money. Vacations should be dedicated in part to a great deal of solitary reading and only a limited part of any vacation should be given to paid work; and we all need some rest and recreation.

The other constructive enterprise is to look for cheap books. Many classics may now be had in such paperback editions as Penguin, Panther, or the American Signet Classics; World's Classics from Oxford University Press, Everyman editions from Dent or the numerous classics published by Nelson and Collins are at least cheaper than some editions.

Most of all, there are second-hand books. If no one buys

new books, authors will starve; but Keats and Conrad cannot starve now.

Real antiquarian bookshops are not, unless they have cheap trays, the student's best bet; old books of interest to scholars and collectors will normally be expensive. Try the yellow pages of the telephone directory to find all the bookshops within reach; in a university town (and many other places) you are likely to find some unappetising, dusty shop that sells old books of little monetary value. That is where you may pick up an armful of various Collected Poems in fussy Victorian editions; faded World's Classics and Everymans, a great many Victorian novels, old 'Mermaid' plays (if you are lucky), pretty little old Temple Classics, ugly old school editions, somebody's Milton that was a Sunday school prize in 1905, three Hardy novels with the bindings ruined by spilled soup, a shabby *Oxford Book of Eighteenth Century Verse* you can afford because twenty-four pages are missing.

The treasures thus acquired will not be exactly a gentleman's library; but they will be a student's books. When there is time, the look of the books can be improved somewhat; marks can perhaps be erased or eradicated; tears patched; missing pages replaced by xeroxes from a library copy; hopelessly damaged bindings veiled in tidy home-made covers. Some day you may afford beautiful books; what you need now are texts to read.

It is worthwhile making friends of these backstreet booksellers; when the bookseller who knows you gets in some fresh boxes of mostly semi-rubbish, he may put aside something likely to interest you. What such a trader needs is quick turnover, and if you come in often enough for forty pence worth, you are a useful regular customer.

Try every charity shop within reach – Oxfam and the like; they usually have a few shelves of books very cheap; you may be lucky, and every tenpence spent there goes to relieve someone far worse off than the poorest British student. There may be an untidy stall in a market. The Students' Union may have some kind of book sale arrangement; departing students offer books for sale; local fund-raising efforts at churches and so on may include bookstalls.

A university town usually has more bookshops than a town with no university; but if your home town, or a town you are visiting, has no university and does have a second-hand bookshop, the very books you need may be the ones it cannot sell and will price low; it is worth looking.

Relatives or older friends may be willing to give you books they do not expect to read again, or let you have books on long loan for your course. In the latter event, play fair with a kind lender who trusts you.

The student should try to own, besides as many literary texts as possible, a few reference books. Every serious student of English should have the best dictionary he can afford, at least one general history of English Literature and a one-volume general encyclopaedia. Out of almost unlimited further possibilities I would recommend as especially useful and interesting: *Brewer's Dictionary of Phrase and Fable; The Oxford Companion to Literature; The Oxford Companion to the Theatre; The Penguin Companion to Literature*, especially, but not exclusively, volume I, *Britain and the Commonwealth; Annals of English Literature* (which shows us at a glance what literary activities were going on at the same time); a classical dictionary: *Everyman's Classical Dictionary* is a good buy, new; but a tolerable earlier classical dictionary is one of the things that with luck a student may pick up cheaply second-hand; a Bible: An old-fashioned Bible with a Concordance and a mass of assorted information may be practically given away and be very useful. Much of the information will be out-dated, but because it is out-dated it may help to explain what earlier writers believed.

To enlarge your own vocabulary for writing essays and papers, and for endless interest and amusement, supplement your dictionary with *Roget's Thesaurus of English Words and Phrases* (now modernized in Everyman and Penguin editions) or, if a wealthy aunt offers you one, the *Reader's Digest Family Word Finder*. The latter is more sumptuous, much more expensive, includes some examples of usage and etymological curiosities, and may occasionally mislead the British student by American usages.

Whatever we cannot possess, we must borrow from libraries or consult in libraries.

The part of the university library that relates to English is normally the place to start. The library almost certainly issues some concise guide, which should be read, as should the library rules, explanatory statements at catalogues and card indexes and any other notices. If freshmen are offered an introductory lecture on library facilities, they should attend it; to save an hour by cutting that lecture may later waste ten hours in assorted perplexities. The first week is the time to learn all the library procedures that concern you and the whereabouts of things you are likely to want, from the *Dictionary of National Biography* and the *Encyclopaedia Britannica* to the place to leave book order slips, from the publications of the *Early English Text Society* to the washroom. You will not want to waste time orienteering on Friday afternoon when you have an essay to finish for Tuesday morning; and the less time library staff have to waste answering the helpless baas of strayed lambs, the more time they have for answering questions that obvious handbooks, maps and notices do not answer.

If you have not previously had experience of more than a school library and the local branch of the public library, you ought to find a university library somewhat awe-inspiring; you may also find it bewildering, intimidating and possibly vexing.

The school library caters largely for obviously immature minds and an important part of its function is to stimulate interest; as an adult student you are supposed to be interested already. The public library dispenses entertainment and general information. The university library is a learned library; its business is to serve specialized education and advanced scholarship.

Such libraries vary in both the degree of general access to the bookstacks they allow and the extent to which (if at all) books may be borrowed for use outside.

Resentment at restrictions is unthinking. A notably gentle, helpful American librarian told me of students who even made noisy scenes when they found they could not take home from the Rare Books Department books of which only a few copies existed, even books of which only one copy was recorded throughout American libraries. Apparently these students felt it as an interference with their democratic rights that they were not allowed to expose irreplaceable volumes to unpredictable risks.

One of the duties of a learned library is to preserve what it may never be able to replace. If Piddlegully County Branch Library loses *Murder on the Orient Express*, that is a pity; it is an entertaining tale; but it is in print, and one such tale is about as soothing as another. If the Verulam Library of the University of New Atlantis loses A. B. Grosart's edition of the *Complete Works of Francis Quarles*, which appeared in 1880–1, the *only* collected edition, this may mean that no future New Atlantis student will ever be able to read more Quarles than he can find in anthologies or selections, and no Quarles research can ever be initiated there. (Yes, microfilms can be bought; but reading them is wearisome and no undergraduate is going to find some crisis of intellectual stimulus browsing among microfilms.)

Acquire, as fast as you can, the habit of reading in a library. It is a good place to work; it is relatively quiet; you have to sit up at a desk; the possibilities for fidgeting and wasting time are limited; reference books are handy when you wish to check something; and in winter you also get free heat and light.

Note-taking in the library is a vital part of adequate study. The possibility of buying xerox copies has, within my lifetime, revolutionized the use of learned libraries: a xerox is infallibly accurate and may save hours. This facility does not, however, render obsolete the practice of taking notes, for two reasons.

Xerox is an excellent buy when a xerox of four pages of double-column small print saves a morning's copying; it becomes very expensive when used on a large scale. Libraries have to observe the laws relating to copyright; an original work is protected during the author's lifetime and for a further fifty years. There is no room here to go into the complex details; but in general a student cannot have a xerox of a whole book, or a whole single poem, that is still in copyright; though learned libraries are allowed to supply xeroxes of articles from periodicals. A student wanting copyright material may have to sign a declaration that he requires it for private study only.

It is not fair to grumble at library staff about this; a librarian cannot be expected to do illegal acts to oblige a student. It is also not fair to grumble about the copyright laws: authors are paid only in proportion to the copies of their books sold; it costs as much to feed an author as to feed a student; and the kind of writing interesting to literary students is very seldom lucrative.

So the student must often copy matter by hand, or take notes in the true sense: summarize, select, paraphrase, tabulate. Manual copying demands real attention, and in copying we begin memorizing; the selection and summarizing and paraphrasing required for taking proper notes demand so much attention as to amount to rigorous self-tuition, a continuous checking of whether we understand what we are reading, if we do it properly.

The library offers more than a collection of English Literature and literary criticism. The history shelves, for instance, may provide explanatory background reading. Encyclopaedias and biographical dictionaries, an atlas or gazetteer, give concisely information such as we may need for full understanding; for close reading of a short extract, a few minutes with the *Oxford English Dictionary*, in its twenty-five huge volumes with two Supplements, may be most rewarding, sometimes surprising, in the light it throws on changed meanings or unexpected implications of words; pictures of the art, architecture or costume of a period may enrich our understanding, including our understanding of actual words and expressions ('this wooden O', 'farthingale', 'conduit', 'wimple'). Purely bibliographical works are used more by researchers than by first-degree students, but it is worth knowing where they are; for instance, the great *New Cambridge Bibliography of English Literature* may well offer the quickest check of some title or date, or reading list for a paper.

The learned periodicals, too, are more for researchers than for undergraduates; but notice them and know where they are; sooner or later a lecturer will recommend an article that is helpful, and it is no use hunting for that book by P. M. Ellay when what you were advised to read was an article in *PMLA*. (If you did not quite catch a reference in full, and cannot catch the lecturer either, try those shelves of bibliographical reference books.)

The university library is unlikely to be the only library available to you. If you are attached to a college, learn quickly how to use any college library and what resources it offers in your subject. The ordinary public library is not to be disdained: it will not have an array of highly specialist scholarly works, but may well provide a standard biography or popular classic when the demand for this in the university library this week suggests twenty castaways wanting to eat the same seagull. The public library at home may be more use for vacation reading than you realize; some city public libraries have good collections of, for instance, modern biography and criticism, and a little country branch may be able to get you a surprisingly good supply of books, especially if you consult the librarian in advance.

There is a huge centralized scheme by which a librarian – never a private person – can procure a book for use in a local library, through the British Library Lending Division, now housed in Yorkshire, expected to have soon one hundred and twenty miles of bookshelves and now dealing with nearly three million requests a year. In certain circumstances this can provide a heaven-sent solution to a problem. However, should you ever have reason to think this procedure might help you, do not expect it to be speedy; do consult the librarian concerned intelligently, giving the fullest possible details of the book you need; and be prepared for some expense; the charge varies from library to library, but has to cover at least the postage.

It may be worth while to find out from local sources of information, or from reference books such as the Library Association's *Libraries in the United Kingdom and the Republic of Ireland*, or *The Libraries, Museums and Art Galleries Year Book*, or the *ASLIB Directory*, whether there is any special library in your university or home town, for instance a former private collection now owned by a trust, that may be useful; such a place may be as under-used as the university library is crowded; you may even sniff out something so little known that the staff are delighted to have a new, appreciative visitor.

It is, however, well to look properly for a book by the most obvious routines before exploring more adventurously or troubling busy librarians. Besides looking at open shelves, make full use of catalogue volumes, card-indexes and bibliographies. Remember that all the works on Donne may not be on the 'Donne' shelf or under 'Donne' in a catalogue: there may be an important chapter in a history of literature, a book on 'Metaphysical Poetry' or a set of essays on seventeenth-century subjects.

And, when perplexed, stop and think. Not long ago, fired by an admirable lecture to read more about Carlyle, I searched in vain for the recommended recent biography by E. N. Campbell until I tried saying it to myself and found Ian Campbell's biography. There is more than one possible spelling for some names: Thompson/Thomson, Philips/Phillips, Stevens/Stephens, Read/Reade/Reed/Reid, Mac/Mc/Mack (the last sometimes all catalogued as if identical) and so on. Foreign names may be spelt with *ö* or *oe*; other alphabets be transliterated in different ways: (Dostoevsky/Dostoieffsky); old libraries may have catalogues in which I and J are treated as identical: Jay, Ibáñez, Jarrold, Inchbald.

Many inexperienced students do not know the common abbreviations used in literary reference books; recent examination papers suggest that some even do not know that 'e.g.' and 'i.e.' do not mean the same; so a list may help. Opinions differ on how abbreviations should be punctuated; but, though I see from an important reference book that many an American learned library now has a 'Chief Cat' on its staff, I have used stops more conservatively.

(That the more unguessable terms are derived from Latin is not meaningless affectation: Latin was the language of scholarship, as of medicine, for several centuries. Further explanations may be found in a good dictionary.)

abr.	abridged, abridgement
app.	appendix
AS	Anglo-Saxon
c.	*circa*, about
cap., ch.	chapter
cc.	chapters
cat.	catalogue (in library institutions, cataloguer)

CHEL	*The Cambridge History of English Literature*
cf., cp.	compare
DNB	*The Dictionary of National Biography*
do.	ditto (the same)
ed., edit.	edited by, *or* edition
EETS	Early English Text Society
e.g.	for example
esp.	especially
et al.	and elsewhere, *or* and others
et seq., et sq., et sqq. (plural)	and the following
fl.	flourished (used of a writer who was known to be alive at a given time, but whose exact birth and death dates are not established)
ib. or ibid.	in the same place
id. or idem	the same (author, usually)
i.e.	that is
ill.	illustration *or* illustrated
in loc.cit., loc.cit, l.c.	in the place cited (i.e. quotation no. 2 comes from the same part of a book as quotation no. 1)
lib. cat.	library catalogue
ME	Middle English
misc.	miscellany
MS., MSS	manuscript, manuscripts
Nat.	National
n.d.	no date, not dated
NED	*New English Dictionary* (now known as OED)
NBL	National Book League
No., no.	number (e.g. of a periodical)
Nos, nos	numbers

n.p.	no place (of publication)
obs.	observation *or* obsolete
OE	Old English
OED	*Oxford English Dictionary*
op. cit.	in the work cited
p., pp., Pp.	page, pages
passim	everywhere, throughout (references to the topic are found throughout the book)
pref.	preface
pseud.	pseudonym
Q. or Qu. or ?	query, question (May it be so? Perhaps)
q.v.; qq.v. (plural)	which see (refer to so-and-so)
sc., scil.	to wit, namely
STS	Scottish Text Society
supp. or suppl.	supplement
tom.	tome, volume
trans.	translation
tr.	translator
u.s.	as above
v., vid.	see
v. infra	see below
v. supra	see above
viz.	namely
vol., vols	volume, volumes

Other common scholarly abbreviations refer to the format of books and to the learned periodicals. *The New Cambridge Bibliography* has a useful list of abbreviations of many of the latter.

A minority of students do not know how to behave in libraries. No one can go far wrong who follows these principles: use the place for study and do not hinder the studies of others; preserve the stock of books; help rather than hinder the staff.

What exactly libraries allow in reading-rooms varies, but spare other readers such things as your Camembert, peppermints, foul feet, unmanageable cough, colds at the most infectious stage, clatter, chatter, sniffs and fidgets. Do not hog a desk for the whole day, or a book in heavy demand, if you are not in fact using them most of the time. Conform faithfully to the boring little routines; otherwise you waste other people's time and, in the long run, your own. Though eating is grand, and cuddling even better, neither is appropriate in a reading-room. In general we should speak as little as possible in a reading-room, but plenty of pianissimo 'Thank-you's tend to keep library staff sweet-tempered.

Finally, vandalism towards library books is even more contemptible than vandalism in, say, the bus shelter, for the student accepted by a university and admitted to its library has not even the thin excuses of someone rejected, ignorant, purposeless, bored.

To steal a book is to rob not only a great institution, but fellow-students and all future generations of students; even when a book can be replaced, every pound spent on replacements is one that could have been spent on new books. To steal one volume of a collection, one number of a periodical, makes the whole set defective.

Occasionally students, in a selfishness that looks almost insane, tear chapters from books or articles from periodicals rather than spend money on xerox or time on copying. Anyone who can thus regard his own temporary convenience as more important than the needs of all other students and scholars has no meaningful concept of the life of the mind and no business in a university.

Not everyone knows that it is harmful to a book to yank it off the shelf by the top of its spine; to mark a place with anything much thicker than a slip of paper; to push the pages back roughly, still worse to bend the binding backwards, in opening the book; to cut uncut pages with anything but a suitable knife used with care. Well, you know now. Common sense tells us to protect borrowed books from rain, damp, overheating, pets, children, coffee or grease; common sense is not as common as one might wish. Anyone who has a genuine accident should own up at once; a librarian may have access to means of remedying the damage, and is likely to be as reasonable about a real mishap as he should be stern about a deliberate act of vandalism.

Anyone who writes in a book not his own is showing an arrogant insensitivity that has almost a flavour of sacrilege. If you are certain, *after real checking*, that a living author has made a mistake of fact, and you feel obliged to do something about it, write a *polite* letter to the author (c/o the publisher), keeping firmly in mind that to write a publishable book is at least a greater achievement than to find one mistake in it. I have seen a written debate between two readers defiling a literary biography in a great library; the level of critical discrimination included such assessments as 'Shit' and 'the whole book is tripe'. The book was doubtless imperfect, having been written by a human being, and a trifle too rhetorical; but it also contained much that was worth reading and once taught me a lot. Those two opinionated boors were not yet fit to enter a library; but even the most temperate, lightly-pencilled 'no' is a mark even the greatest scholar has no right to make in a book that is not his own.

7

EDITIONS

'He must be a clever chap; I'd no idea what he was talking about.'

'Yes, he is a very distinguished scholar in bibliography and textual criticism.'

'Then why doesn't he wear his dog-collar?'

Conversation in a pub near a library

Not all printed 'copies' of a literary work contain exactly the same words. For instance, there is a mass of scholarship comparing and discussing the First Folio of Shakespeare's works (1623) and the less 'official' but sometimes important 'Good' and 'Bad' Quartos of many of the plays, and the numerous subsequent editions. An hour or two spent with a volume of a variorum edition (i.e. one that shows various possible readings, usually with notes) may be instructive and surprising to the novice. Shakespeare had little to do with the copying of his own plays; but even when authors have corrected their own proofs, as for instance Byron sometimes did, and as is today normal practice, different versions may come into existence.

Some of the most exact, exhausting and honourable scholarship is that which aims at establishing the text probably nearest to the author's intention (which may itself have changed over some years). The sciences involved – descriptive bibliography, textual criticism and (for manuscripts) palaeography – are highly specialized, detailed and difficult, and are likely to make only a token appearance, if that, in undergraduate studies.

Undergraduates do, however, need to know that there are such things as good and bad editions.

The student has to look for cheap books; not only the student, but the established scholar, even the buyer for the library, is restricted not only by what he can afford, but by what he can find. (A multi-millionaire cannot tell his friendly neighbourhood bookseller to send round a copy of Donne's *Ignatius his Conclave* (1611) this afternoon.) For a broad, general, impressionistic reading, say of *The Way We Live Now* as an example of Trollope's art, or Fletcher's *Philaster* as part of a programme of sampling Shakespeare's contemporaries, at undergraduate level, any text we can get hold of will serve. More useful background knowledge is gained by reading six plays than by reading only two because so much time was spent finding the very best text.

For closer work it is an advantage to use the best possible edition.

Actual *set books*, prescribed for very detailed study, on which there may, for instance, be context questions ('gobbets') in examinations, and from which the student needs to be able to quote abundantly, should always be studied in the edition recommended in the syllabus or by the appropriate lecturer. If the edition is recommended, this is for sensible reasons – it is the best text, it incorporates findings of recent research, its footnotes are the most useful, it has an especially valuable critical introduction – possibly, even, another edition is notoriously bad. Examiners will, fairly, assume you have used the prescribed edition, having been told to do so.

Suppose the prescribed edition is distressingly expensive, or not available. Do not assume it is not available because it is not on a bookshop shelf; it may just be out of stock, and can be ordered. If it has really gone out of print, tell someone in authority, as this may affect the next meeting to discuss the syllabus. But if you truly cannot find the book, or the price of it, do as I sometimes did: take your inferior edition to the library and compare it all with the prescribed edition; copy into your margins notes of important differences. If you need a great deal of matter, interleave your copy with thin paper and fill that with notes; and make notes on any useful extra critical or background matter.

This is a chore; but after doggedly plodding through it you will already have a far better knowledge of your set text than the lucky student who has just written his name in his gleaming new copy of the prescribed edition with his gold fountain-pen.

Notice, by the way, that in different editions there may be discrepancies in the numbering of lines, which may affect the identifying of footnotes or quotations.

True scholarly editing, aimed at the reader at undergraduate level or above, involves at least one of three possible functions: to provide a carefully considered *text*, so that the reader is given, as nearly as possible, what the writer intended; to add to the text some helpful *background material*, such as a biographical sketch, a placing of the work in its context in general history and the history of literature, an exposition of particular literary traditions (or innovations) and techniques involved, and explanation of the textual history and problems; a general critical appraisal: and to provide *explanations* of the more perplexing points in the text, by means of *annotations* – footnotes, marginal notes, or notes at the back of the book – or perhaps a more general *glossary*.

Some editions of well-known works are intended chiefly for school use. The best of these are scholarly as far as they go. Some editions cater for a wide range of readers, e.g. the invaluable 'Arden' editions of Shakespeare. Some are not intended for serious literary students, who should avoid them if possible.

A few specific warnings may be helpful.

Many editors, probably most of those whose work undergraduates use, have modernized the spelling and punctuation of texts later than Middle English. This is usually harmless, so long as the work is being studied as literature and not as a philological specimen; and may even be helpful, as an odd archaic appearance can distract the novice. This, for instance, is First Folio spelling:[1]

Cleo. Giue me my Robe, put on my Crowne, I haue
Immortall longings in me. Now no more
The iuyce of Egypts Grape ſhall moyſt this lip.
Yare, yare, good *Iras*; quicke: Me thinkes I heare
Anthony call: I ſee him rowſe himſelfe
To praiſe my Noble Act.

Students should, however, be careful not to draw minor critical conclusions from modernized spellings; and it is often the good student who, capable of noticing details, makes an intelligent but misplaced comment on punctuation which is not an author's but a modern editor's. Editors even occasionally modernize vocabulary, which may be helpful to a school child, but is misleading for the serious student.

Weak students sometimes make bad blunders; a fairly common one is to read, quote and appraise Nevill Coghill's *translation* of *The Canterbury Tales*[2] – a delightful volume that brings some of that entertaining work within the reach of the non-specialist reader – as if it really were Chaucer's *Canterbury Tales*.

An old book (or set of books) may be an edition that is now badly out-of-date. It may contain items now dropped from the canon: for instance, the poem '*Absence, hear thou my protestation . . .*' used to be attributed to Donne, was later attributed to John Hoskyns and later still was regarded as of 'doubtful authorship'.[3] Or it may lack items subsequently found and possibly important: any 'set' of Jane Austen published before R. W. Chapman's collected edition of 1954 lacks several amusing juvenilia such as *Love and Freindship* and *Lesley Castle* and the draft of *Sanditon*, on which she was working just before she died; these are interesting both for themselves and for the light they throw on her artistic development. Footnotes may include assertions that have since been disproved; a Victorian introduction may make critical judgments that seem to most present-day readers prudish, timorous or pompous.

A much more serious danger to the unwary is the abridged version; and it is from such versions that many of us receive the beginnings of our literary education. Some version of 'Stories of the Round Table' read in school is nothing like Malory's *Le Morte D'Arthur*; a school *Robinson Crusoe* falsifies (advantageously from some points of view) the proportions of Defoe's more uneven complete text; the little *Don Quixote* I remember from my schooldays gives no idea of the massiveness of Cervantes's masterpiece. Distrust: school texts, checking how far they are shortened versions; anything marked 'Abridged'; texts prepared for foreigners studying English, such as editions once made for imperial India; anything called a 'condensed novel' or the like; 'The Book of the Film'; and a 'Selection' may at times be useful, but for other purposes be misleading and insufficient.

Much abridgement is a form of simplification, usually for the purpose of introducing some classic to children. Sheer economics and pedagogic realism play a part. If a class of thirty-five needs a set of books, a complete translation of *Don Quixote* is simply a financial impossibility, before other considerations arise; and people who are desperately trying to coax children towards some relish for reading are quite rightly more concerned with pulling out plums than with presenting an intact pie (with, possibly, some indigestible pastry anyway). A university student, however, needs at least to know the true state of the pie, even if he cannot eat it.

Editions intended for school use, and some other editions, are often marred by bowdlerization; silly and immature though it is to sniff like a smut-hound through literature, the scholar more than the smut-hound may resent excisions that distort an artistic intention. Expurgators with the classroom in mind may have better motives than most censors: one child spurred by mingled prurience and spite can find means in some sexual allusion to disrupt a lesson hastily, perhaps upsetting some more sensitive, shy or sheltered children as well as the hapless teacher; we are justified in keeping explosives out of the hands of terrorists. But a serious scientist studying nitrogen compounds needs to know about nitrocellulose and nitroglycerine.

A school edition of *Gulliver's Travels* may give a completely misleading impression of that masterpiece of black humour; it may emphasize the amusing fantasy of Gulliver's pleasanter adventures among six-inch and sixty-foot people, or the good nature of rational horses, playing down or omitting altogether the savage ironies, the loathing of human folly and the ugly images; we may find the cat whose purring was 'like that of a dozen stocking-weavers at work'[4] but not the repelled-fascinated look at the giantess's nipple.[5] Without the spasms of obtrusive repugnance, we have an interesting book; but we do not have Swift. I have seen a school edition of *The Wife of Bath's Prologue and Tale*, in which there was much helpful additional matter; but the most forthright sexual allusions were omitted; this falsifies Chaucer's characterization, since the frank, unashamed appetites of the Wife of Bath are an important aspect of her. School editions of Shakespeare plays are often similarly mutilated. Worse, such editions may contain genteel emendations; Chaucer wrote of his Somnour:[6]

> As hoot he was, and lecherous, as a sparwe.

I first met him in a school edition in which

> And quick he was, and chirped as a sparwe

and I even learned this wrong version by heart for an examination. Old editions may also delete or soften bits of profanity, sometimes because of shifts in the relevant laws.

We should, however, be thankful that in Britain there is little censorship or expurgation except for juveniles. Several powerful religious bodies have tried, and in some places still try, to suppress literary works as contrary to their doctrines; the Nazis literally burned books of outstanding cultural value; in a communist state which – to protect my informant – I do not identify, a large number of books in the greatest national library were destroyed as 'no longer of interest now we have had the Revolution'. I believe that all scholars and students should in some way, however trifling, be committed to the concept and positive defence of intellectual freedom everywhere.

Not even intellectual freedom at the British level protects us against commonplace human fallibility; many editions of literary texts are poor through mere slovenliness. Not long ago a specialist scholar had reason to check through a Jane Austen novel as published by a respectable firm with an honourable record of educational and cultural service; and found over two thousand textual faults, many slight, but some serious, such as the omission of whole sentences. This will not be a unique case.

Many publishers also have a 'house style', to which everything they print must conform; this may involve alterations of spelling, punctuation, usage of such things as italics and capitals, and so slightly modify literary texts; though these details are not likely to affect the undergraduate reader.

Though such a reader is not much concerned with textual problems, and it is better to read a very imperfect text, even an abridgement, than to know nothing of a book, even the youngest student sometimes has to know a little about textual variants if studying a work that exists in versions so different that the differences may be significant in literary history or critical appraisal.

There are basically two versions of *The Ancient Mariner*: the text in *Lyrical Ballads*, 1798, and that in the second edition of *Lyrical Ballads*, 1800. Many of the changes remove rather pointless archaisms (e.g. 'an' becomes 'if' and 'ne' becomes 'nor') but here, for instance, the whole style of a verse is strengthened:[7]

1798 Are those *her* naked ribs, which fleck'd
The sun that did behind them peer?
And are those two all, all the crew,
That woman and her fleshless Pheere?

1800 Are those *her* ribs, through which the Sun
Did peer, as through a grate?
And are those two all, all her crew,
That Woman, and her Mate?

If we are examining poetic technique, differences like this do matter.

The difference between the original 1805 version of *The Prelude* and the revised 1850 version, neither of which was published in Wordsworth's lifetime, are of real critical importance and the changes themselves contribute to our knowledge of 'the growth of a poet's mind', with not only some obvious improvements of wording, but some stylistic and philosophical modifications that are more controversial, and the significant mutilation of the story of Vaudracour and Julia in the ninth book. The student whose course includes a close study of *The Prelude* will have to examine these differences in some detail; but any student reading *The Prelude* at all (and a course in English Literature that did not demand such a reading must surely be deficient) should know which version he is reading, and why.

Editions of *The Ambassadors* differ considerably, even with additional chapters and transposition of the order of chapters; the edition of 1909 is now reckoned the best text.[8] *Tender is the Night* exists in virtually two versions, including, again, transposings of the order.[9] Herman Melville's *Billy Budd* was published posthumously from a much revised manuscript which had not been made into a fair copy, and left some of Melville's intentions doubtful; markedly different printed texts appeared in 1924, 1928, 1946 and 1948, and a perhaps definitive version with full notes only in 1962. W. H. Auden several times revised his poems after they had been published, so that more than one 'authentic' text exists.

Until the Copyright Act of 1842[10] and even, less frequently, later, literary piracy, the unauthorized publishing of works that might sell well, was a problem; naturally such furtive printings, especially of plays, were often very unreliable texts. Recent research in Moscow discovered that Tolstoy's wife, fair-copying his manuscript of *War and Peace*, made 1,885 errors which went into all editions until recently, and have presumably affected all translations. D. H. Lawrence died in 1930, but it is unlikely that the Cambridge University Press will complete its definitive edition of his works, with massive corrections of all printed texts previously available, until 1990. These are only samples of the kinds of difficulties that often arise.

Most students find a prefatory 'Note on the Text' dull reading; but it may be prudent reading; before venturing a critical judgment, we should have some idea of how confidently we may assume that we are reading what the author wrote.

It is possible to fuss too much; no undergraduate has the technical knowledge to understand fully, let alone solve, textual problems; most texts will serve most general purposes. However, though the undergraduate has to take much textual information on trust, he can and should understand that there are such problems; that scholarly editing, even authorial proof-reading as standard practice, came comparatively late in literary history; and that few people can copy correctly, as indeed his own notes may occasionally demonstrate. What can happen to manuscripts is dramatized in *Chaucers Wordes unto Adam, his Owne Scriveyn*; and the fallibility of printed books is exemplified by the edition of the Bible, *circa* 1702, in which David, usually persecuted by princes, laments in Psalm 119, verse 161 that 'Printers have persecuted me without cause . . .'. We can only try to prefer the more accurate to the less accurate.

8

THE AUTHOR AND THE CRITICS

Study is like the heaven's glorious sun,
That will not be deep-search'd with saucy looks;
Small have continual plodders ever won
Save base authority from others' books.
Shakespeare, *Love's Labour's Lost*[1]

Perhaps volumes of literary criticism should bear the reminder:

> **INTELLECTUAL HEALTH WARNING:**
> **RELIANCE ON CRITICS CAN SERIOUSLY**
> **DAMAGE YOUR INTEGRITY.**

The danger is not so much in the critic, as in the student, who may be tempted to use the critic's book as it was not meant to be used. A map is very useful to someone who is exploring a beautiful city, especially if he finds its geography hard to grasp and keeps losing his way; but the most diligent study of a map – even to the point that we can reproduce large parts of it – tells us nothing about the city comparable to the experience of walking round it and looking at all the sights.

I have already quoted Johnson's eminently sound advice on the right order of literary studies. His is a recipe for both enjoyment and learning.

There is a recipe for intellectual dishonesty which may also be relied upon to neutralize most of the savour of the study programme. It is the attempt to 'get the gen on' Jonson while reading as few of Jonson's plays as possible, to 'bone up on' Byron without even a desultory reading of *Don Juan*; to collect things to write about writers, in essays or examination papers, instead of investigating what the writers wrote. The short answer is, 'Grrrr! if you did not want to read a lot of literature, why did you embark on this course?'

To be fair, this practice of reading books about books in order to quote them, as a substitute for reading the books themselves, does not always begin in conscious intellectual dishonesty; it is more likely to arise from being short of time, or finding an unfamiliar genre of literature bewildering, or even a just and creditable awareness of one's own incompetence: but it is a procedure for first failing to develop more competence, then ceasing to realize the truth of one's continuing incompetence, and finally acquiring a conceited generalizing glibness about literature that may infect other aspects of one's life until it spoils a whole personality. What may well have begun partly in a genuine humility, though a humility nervous and defeatist rather than generous and honest, may end in that particularly senseless and ludicrous kind of pride we can so easily take in our inadequacies.

We need to make a distinction between *preliminary* reading and *premature* reading.

A biography of the writer, that places him and his works in their historical setting and gives us an idea of the experiences that may have most affected him, may enable us to start reading his book with more intelligent expectations. A good book on the history of the period (not necessarily even mentioning our writer) may be a great help. Before reading *Samson Agonistes* we do well to familiarize ourselves with the original story of Samson in the Book of Judges; we cannot discuss 'how Milton treats his material' until we have seen the material he is treating; and to read a few Greek tragedies, even in translation, even hurriedly, is better than nothing to give us some idea of the literary tradition in which Milton, a man of deep classical culture, was working, when he wrote a poetic drama with a Chorus, two characters on stage at a time and all action off stage. We must read Richardson's *Pamela*, or at least a substantial sample of it, before we can fully appreciate Fielding's *Joseph Andrews*; without knowing something of the *Iliad*, or at least about epic tradition, we shall not understand what Pope is doing in *The Dunciad*. Some preliminary reading is essential, above all, when we are reading medieval literature; we are almost helpless if we know nothing of the obsolete picture of the cosmos; the religious assumptions; the patterns of feudal society; the conventions of chivalry and courtly love; various concepts found in, for instance, astrology, alchemy, bestiaries, herbals; the state of geographical and historical knowledge; even some unfamiliar details of everyday life.

There are many books written as *introductions* to the work of various authors, intended to whet our appetite for the works themselves, give us some background information, place the author in literary history. Among such books are Macmillan's extensive series of 'English Men of Letters', such as Harold Nicolson's *Swinburne* (1926) or J. B. Priestley's *Thomas Love Peacock* (1927) – an old series still eminently helpful, stimulating interest and usually written in lucid sentences not clotted with pretentious jargon. Another admirable series is the very wide-ranging collection of pamphlets, 'Writers and their Work', published by Longmans, for the British Council and the National Book League: Bonamy Dobrée on *Dryden*; Kathleen Raine on *Coleridge* and so on – offering, by 1978, a total of two hundred and sixty-three useful introductory essays, each with its reading-list.

An introduction of this kind can do us nothing but good – provided only that we read it as helpful first words on the subject and never as the last word on the subject. Even the unappetising little compendia sold as aids to passing the school-leaving examinations on set books are not useless; they often contain relevant facts and sensible explanations; but they must be seen as minor diet supplements, not as magic pills.

Any normally intelligent student can soon tell from reading a few pages whether a book is an introductory volume that will provide some background and perspective, or whether it has been written for readers who have already read the works being discussed. To read a book of the latter type before reading the works themselves is not useful *preliminary* reading, but imprudent – and intellectually dishonest – *premature* reading, likely to confuse and oppress the sincere student, or give the less sincere student material for phoney judgments at which he has never truly arrived.

Any student of literature has the right to quote any critic in an exercise or discussion; but one of the most important parts of such a quotation is a pair of quotation marks, or some acknowledgment such as, 'as M. R. Ridley says in *Keats's Craftsmanship* . . .'. When we mention, approve or refute someone else's opinion, we are joining in a wonderful, endless debate; when we try to palm it off as our own, we are merely cheating and evading. Premature reading of critical studies leads to a lot of cheating.

While there are exceptions to almost all rules, there is an order of study which in general works better than others, both for outward academic success and our own personal development.

1 Do a bit of preliminary reading if you can find something appropriate; collecting any information that may be useful, but never deluding yourself that this informatory equipment is any substitute for literature itself. A signpost is not a walk.

2 Either first, or after a little preliminary reading, read the prescribed work of literature. This is the most essential of all

possible study processes. Critics may cover miles of paper discussing an author's exact intentions; but one thing we do know about the author's intentions is this: he intended to be read.

Your own instant, untutored, probably in part immature as well as uninformed, subjective impression of the literary work is not the end; you have a lot to learn; but it is the only sound beginning. At least now you have had a genuine experience of the book. In discussion, you have something to discuss; in reading criticism, you at least know what the critic is discussing. None of us can totally trust his own judgment, or anyone else's, about anything; but until we have our own judgment, however provisional, tentative and qualified it ought to be, we cannot proceed further with any reason, reality or integrity.

Moreover, if that first reading is too much motivated by a quest for something to say about the work in an essay, it is very far from the author's intention. All study of great literature should be rather humble; how many of us can write even one-tenth as well as, let us say, Wilkie Collins or Christina Rossetti, let alone Dickens or Donne? We shall never become wise by being too eager to seem clever.

The Estonian poet Mart Raud (b. 1903) offers an illuminating image:[2]

> Opening a book is like opening a door to the knock
> Of one who has come from afar; and the heart gives a token
> By its leap and its louder beat at the moment of opening
> the book
> Of another knock; it is also yourself you must open.

This opening of ourselves, this readiness to receive experience, is essential.

3 The next stage in literary studies is to check, as far as possible, that our experience conforms to the author's intention. Without receptivity and sensitivity we might as well

abandon the study of literature; but these gracious virtues do not automatically endow us with universal knowledge and infallible comprehension.

So next come annotations, explaining puzzling expressions, identifying allusions we may have missed, pointing out implications we did not see; single glosses, for early literature even whole glossaries, explaining individual words; explanatory introductions and appendices; perhaps a *Companion to* . . . or a *Notes on* . . .; perhaps, too, a class for close examination of the text. And do not forget that your ordinary dictionary may often throw light on a passage you are not sure you understand.

4 No book that is worth reading yields up all it has to give at a first reading. This is not merely because, at least for old or difficult literature, we may need annotations. It is also because the attentiveness of even a fairly well-trained mind is imperfect, sporadic, somewhat self-deceiving. It is also because our own capacity for experiencing at one time is limited here, as it is with regard to everything else; we are in a perpetual state of never quite catching up with all that is happening to us. Skills, whether for pastry or music, karate or driving, come only with practice; and all those receptivities that may seem passive – reading, listening to music, looking at paintings, delighting in other people or in the world of nature – are really unobtrusive skills. To gain something like the full experience of a great work of literature, we need in some sense to practise experiencing that particular book. Which means, of course, re-reading.

We often cannot re-read as much as, ideally, we should, because we do not have time. We can only try to re-read as much as we can.

5 It is only when you are well acquainted with the book itself, and have taken all reasonable steps to ensure that you know the meaning of the words, that you should read much criticism that goes beyond the obviously factual and ex-

planatory to more probing analyses, complex insights and elaborate theories.

On most of the authors you are likely to study in detail, there are substantial bodies of critical writings, and you will not be able to read them all. Selection should not be more random than can be helped, though it is bound to depend partly on such non-literary factors as what is available in the libraries to which you have access, and how much time you can give to such reading. Bibliographies may be helpful; lecturers are not infallible, but their recommendations are more likely to be wisely selective than shutting your eyes and sticking a pin into a list. (Besides, lecturers are likely to recommend books helpful in relation to the examinations of the university.)

When reading critical works, or even annotated editions, notice who is saying what, and when. An experienced tutor once observed drily to me that '*It says at the back* is an impersonal verb in frequent use.' *It* does not say at the back, any more than *they* should repair the hole in the pavement; authorities have names and status. Refer to an edition of the text by the name of the editor or at least the series; Johnson's Shakespeare, the Arden *Twelfth Night*, H. Harvey Wood's *Poems and Fables of Robert Henryson*. Even a dictionary has a name: *Chambers Twentieth Century Dictionary*, *The Shorter Oxford English Dictionary*. (Most of the time we may abbreviate: Johnson, Arden, Wood, Chambers, *SOED*.).

It is by no means unheard of for a student to read, and regurgitate, a critical work without remembering who wrote it or ever having looked at the date. When making notes on any critical work, first put down clearly the title, the author, the date, the place, and perhaps the publisher – the last may help if you are looking for the work again some day.

While there cannot be any wholly reliable quick formula for judging the worth of a critical work (and critics on critics disagree, as critics on primary authors do), there are some reasonable indications. There are scholars, about whom other scholars will know, whose work has earned general respect in certain fields: R. W. Chapman or Mary Lascelles on Jane Austen; Helen Darbishire or E. de Selincourt on Wordsworth; Leslie A. Marchand on Byron; Eric Partridge on the history of slang, Fredson Bowers on modern textual criticism. Conspicuousness in the Press or on television is not a trustworthy measure of scholarly merit, though it is also not in the least an inevitable proof of phoneyness.

A good critical study that appeared in 1895 may contain matter that has been rendered obsolete by later research, though if it was ever any good it is sure to contain something still worth reading. Other things being equal, a book by someone in the English faculty of a major university, or published by a reputable university press, is likely to be of more worth than one by a village schoolmaster published by a small provincial press that may even be a 'vanity publication'; but there will be exceptions; dedicated scholars are not all employed by universities. (Be a little wary of little-known American universities and their presses. Some of the finest scholarship in English Literature is being done in North America, but there are a few backwoods universities of deservedly poor reputation.)

Approach with some caution anything that is presented in some way as a sensational book, a startling new interpretation, and so forth. It is by no means necessarily false or foolish because of this; but there is at least more risk that it may be heavily biased, or straining for effect, or perversely rather than illuminatingly original, than something quieter, more solid and perhaps less immediately exciting. And, though biography clearly has something to contribute to our understanding of literature, we should also remember that, for instance, spicy details of Byron's sex-life, while interesting to most of us, contribute no fresh insight into his poetic method; the most they may do is throw some light on the genesis of a particular poem. To whom a love poem was written, for instance, may be interesting. You may sometimes enjoy reading a juicy scandal that can now hurt nobody; so do I; but scandal is not literary criticism.

Do not, incidentally, take too seriously articles on literary subjects in the popular Press; reviews may well point to a book you ought to read; an article may suggest some line worth following up; but much journalism lies almost at the antipodes of scholarship. At worst it is wilfully biased and shamelessly negligent; but at best it has to be done in haste and confined to scanty space, so it is likely to be distorted by selectiveness if not by worse dishonesties.

Even when we have before us a critical work that is clearly by a serious, reputable scholar, we need to approach it with a certain ambivalence; an uncomfortable mixture of deep respect and discreet mistrust.

A devoted scholar who has written something on, say, *Paradise Lost*, quite certainly knows more about *Paradise Lost*, epic, Milton, seventeenth-century background, and critical method, than any student. He has made sustained and complex intellectual efforts such as no undergraduate has yet made. He probably has a broader general culture and a more powerful intellect than most of us. All who think the life of the mind is a life worth leading should respect him. He deserves our admiration, our gratitude, and what he probably most wants of us – our attentive reading.

On the other hand, the greatest scholar still has only one human mind and only his own particular pattern of emotions; so he shares our universal human capacity for narrowness, obsessions, unfair selections, distortions of fact to fit opinion, and even mistakes on facts. Some of us look at things a great deal more carefully than others; but the closest observers, like all the rest of us, are still wearing tinted spectacles.

The inference from this need not be to despair of literary criticism, or of intellectual effort in general. A perpetual passing on of imperfect knowledge is inevitably the only way in which human knowledge develops and we have come out of the Stone Age. But the student should as far as possible avoid depending on any *one* critic. He should not become a disciple of one particular critic nor claim to be aligned with one school of criticism. It is possible to have tremendous respect for a great mind without making an unconditional surrender to it as an infallible guide.

Suppose the work under discussion is *Paradise Lost*: then you will be well advised to read E. M. W. Tillyard's *Milton* (1930), C. S. Lewis's *A Preface to Paradise Lost* (1942) and William Empson's *Milton's God* (1965).

Any one of the three will teach you a great deal that you did not know before; any one of them is in many ways a model of a critical work – in its freedom from waffle or unnecessary jargon; in its literate, lucid style; in the stock of knowledge behind it; in the immediately obvious quality of mind shown in it; compared with most of us, all three scholars wrote as princes of intellect. However, a reading of these three fine studies is likely to leave you, not contentedly confident of knowing all the answers, but in the much more healthy but less comfortable state of feeling you now know a great deal more about Milton than when you started, yet never will know the right answer. For Lewis's treatment is very much coloured by his Christian convictions (Protestant – had he been Roman Catholic or Russian Orthodox the colouring would have been of a different shade); Empson's is at least as much coloured by his feelings of repugnance towards what he understands as traditional Christianity; while Tillyard, probably the most balanced of the three and the most useful to a student, has more of a psychological approach. Just to complicate things further, if it is the 1956 edition of *Milton* that you use, you will find that in it Tillyard admits that between 1930 and 1956 he had changed his own opinions on an important part of his view on *Paradise Lost*! Then, just to tie an extra knot, you may like to read A. J. Waldock's *Paradise Lost and its Critics* (1947) in which an American scholar makes some criticisms of Lewis and Tillyard; or to examine F. R. Leavis's view of Milton in *Revaluation* (1936).

After which, you will probably want to relax; and you might do so with that wise, entertaining little book by Frederick C. Crews, *The Pooh Perplex* (1964), in which, by parodying, and extrapolating into hilarious absurdities, twelve possible critical approaches (all real and recognizable) as applied to *Winnie-the-Pooh*, Professor Crews (himself an English specialist) shows how a style can become mannerism, and a useful critical concept can become such an obsession that it becomes not tinted spectacles – we can still see something through those – but blinkers.

After several experiences of seeking help from critics, and finding such alarming differences among specialists who have obviously really given close attention to the subject, you may find yourself wearily quoting Pope: 'Who shall decide, when Doctors disagree?'[3] and perhaps, as a student, wanting to emend his next line to something like 'So what's the value of a PhD?' If scholars of deep learning and mature experience come to very different conclusions, how can anyone expect an inexperienced student to produce the right answer?

The immediate, practical, comforting reply is that the student is expected to produce a *right* answer only on a question of fact. Scholars do not dispute that Milton died in 1674; that Giles Fletcher wrote *Christs Victorie and Triumph* and his brother Phineas wrote *The Purple Island*; or that *The Eve of Saint Agnes* is written in Spenserian stanzas. On questions of taste, interpretation, appraisal, the student is asked to do no more than that which, on a higher level, is the utmost the great scholar can do: to present a *reasonable* answer in which opinion is supported by relevant, organized evidence. For examination purposes the function of the evidence is to demonstrate that the student has studied the book properly; but the professor has only mastered, matured and refined intellectual procedures that the very greenest student is (or should be) using.

It would be very odd if literary critics did not disagree; this would make literary criticism different from all other fields of human experience and knowledge. Two reasonable people who know a third person quite well may react to him in markedly differing ways and assess his personality very differently. This need not mean that one of the two is stupid, unobservant or insensitive, only that the scope for different assessments is enormous: any developed human being is a creature of awe-inspiring complexity, ambiguities, inner contradictions, incommunicable experiences, innumerable factors that have made him what he is. And the two who have different ideas of his personality are two other beings just as complicated, mysterious and unique. No wonder they differ.

If real people were not so mysterious, we should not have literature, let alone literary criticism; literature consists mostly of fragmentary presentations of our human experience, probably inconceivable as a whole. I suspect that critics differ most conspicuously when treating of those writers whose work approximates most to something of the rich, perplexing multiplicity of life: the first examples that come to my mind are Shakespeare, Wordsworth, Dickens and D. H. Lawrence. Jonson's men in their humours are not as controversial as Shakespeare's or sometimes Browning's characters. Sometimes an author presents an experience of such layered complexity that each of several critics will respond mainly to one layer.

About the best we can do is: distrust dismissiveness, dogmatism and discipleship; attempt appreciation, accuracy and adaptability. Then we may at least hope to end a little wiser than we began.

9

SOME WAYS OF MISJUDGING

> The quickest of us walk about well wadded with stupidity.
>
> George Eliot, *Middlemarch*[1]

Since it is part of our inescapable, essential human situation that we live in a perpetual state of fragmentary experience and understanding, it follows that we frequently misunderstand.

No effort and no rigorous training can equip us with infallible judgment; but, conversely, even the smallest effort and training will bring about some improvement.

We can at least try to read and appraise as befits civilized, honest, self-critical, informed, tolerant, modest adults who are lucky enough to live in a society comparatively free and eclectic. Many tyrannies of assorted colours compel ignorance and, even at university level, enforce a narrow outlook that is virtually anti-education. We should not forget to be thankful that we are allowed access to the whole range of our own literature, foreign literatures, and interpretations of literature.

Let us at least admit our own fallibility. Gonville and Caius College, Cambridge, offers a perpetual reminder of what real learning means. Dr Caius's intention was that students should first enter college through the Gate of Humility and eventually go to the degree ceremony through the Gate of Honour. The college has since been somewhat rearranged, but the sixteenth-century allegory remains true: the Gate of

Humility is the way in to sound knowledge.

We can remember that we are likely to forget and mistake; so we can look things up, check, verify. We can admit that in the past year we changed our minds about several things – so we might change our minds about something next year; thus it is prudent as well as polite to avoid a dismissive, dogmatic, aggressive tone in essay or discussion, and to imply by our manner that our conclusions are provisional. We can try to guard against choosing, or, worse, distorting, facts to fit some cherished critical principle or other opinion; if we have to mishandle facts to fit the theory, it is time to ask ourselves whether the theory is as universally applicable as we want to think. Before loudly asserting, 'This is rubbish!' we can pause to ask ourselves if what we should be doing is softly admitting, 'I'm a rabbit . . .'. Since we shall never be rid of them, we can at least try always to bear in mind: the limitations of our experience; the endless intrusiveness of our often very petty subjectivities; the distorting fact that, reading, we are inevitably at a different point in time from that of the writer, so our sympathies cannot be perfect. To understand how likely we are to misunderstand is the first small step towards some kind of passable understanding.

Though we are luckier than most of us can ever realize, to live in a community that allows a full range of reading and debate, we also live in a community in which advertising and publicity are too important, and this probably reinforces the hold on us, always powerful, of mere fashion. Fashion tends to narrow taste; and as most of us have at least some craving to be thought smart, in, with-it, and those who are most modish are often better than others at intimidating sneers, we are apt to follow fashions like a lot of sheep. Sheep provide the best known traditional metaphor for erring and straying. Hardy describes this herd mentality:[2]

When Wordsworth was enthroned they carried pocket

> copies; and when Shelley was belittled they allowed him to grow dusty on their shelves.

– thus demonstrating that they had never really appreciated Shelley.

My own real personal preference, resulting from at least a genuine individual response to my own genuine exploration, has at least some authenticity; it may be what many would regard as an immature taste, but all taste has to start by being immature; or wrongheaded, but every judgment will seem wrongheaded to someone else somewhere; our own real experiences are the stuff of our growth. A skilled tutor, a fine lecture, an intelligently appreciative book, may guide us to some fuller response, show us the merit in something that previously left us unimpressed. That too is real; that too is part of our growth. An unexamined conformity, an easy following of fashion, making a cult where it is expected of us and (more probably and more damagingly) sneering where we are supposed to, is insincere, pretentious, snobbish, cowardly; and helps to prevent our growth.

Kipling spent some time in the critical doghouse because of his crudities and the imperialist and racist implications in some of his work – though opinions on Kipling's opinions sometimes involved unrepresentative quotation or careless reading; today his real craftsmanship can be appreciated and we can admire the masterly impersonations of *McAndrew's Hymn* or *That Day*, or see that, though some phrasing is now unacceptable, *Fuzzy-Wuzzy* or *Gunga Din* are at least as anti-racist as racist; or savour Kipling's sympathies and ironies and even insights in presenting experiences not often treated in poetry.

When I was adolescent, it was fashionable to turn up the nose at almost everything Victorian; it took moral courage to respect Tennyson; today Victorian studies are booming and it is realized that Tennyson's huge varied garden contains much more than just that 'white flower of a blameless life'. Donne, however, was coming into his own after a long period of comparative eclipse. Johnson's essay on Cowley[3] shows how even one of the finest and most independent minds of the eighteenth century found the Metaphysicals too alien for full sympathetic appreciation, and today Johnson's statement that Cowley was 'undoubtedly the best' of 'that race' seems grotesquely inept. (And a bookseller's idea of a Collection of the English Poets could then include Broome, Mallet, Sprat and Walsh while excluding Chaucer, Spenser, Donne, and Herbert.)

Johnson could no more help being the child of a time than can any one of us; but he had too big a mind to be a cheaply modish sneerer, and that same essay on Cowley includes his salutary reminder, 'Wit, like all other things subject by their nature to the choice of man, has its changes and fashions, and at different times takes different forms.'

Changes in literary practice have more significance than superficial critical fashions, and are indeed part of the proper matter of our studies: for example, in the sixteenth century the influence of Petrarch brought the sonnet to England; English poets developed the specifically English, 'Shakespearian' sonnet form; virtually all serious poets wrote some sonnets, until the form was, for the time being, exhausted, and virtually disappeared in the eighteenth century, to be revived by the early Romantics with a different tone and content. Heroic couplets are hardly ever used today. The nineteenth and twentieth centuries have seen enormous and numerous developments of the novel as a major literary art form. Such shifts of the actual creative impetus and direction of technical effort are aspects of true artistic development; writers must not go on imitating or repeating; but when they find that they have for the present come to an end of what can be done with sonnets, this does not discredit the good sonnets already in existence.

Moreover, the pioneer writer, who turns the creative energy into a new channel, is frequently the one who is kicked by the herd of sheep. If we need help on resisting the temptation to 'get with it' instead of truly getting within it, a useful prophylactic is E. E. Kellett's *The Whirligig of Taste* (1929); but the best prophylactic of all is wide, receptive, attentive reading.

Another mistaken approach to literature is to pursue stuffed owls with such disproportionate eagerness that we fail to see angels with immense and luminous wings.

In 1930 D. B. Wyndham Lewis and Charles Lee brought out an admittedly amusing 'Anthology of Bad Verse' called *The Stuffed Owl*, the title being taken from a sonnet in which Wordsworth tells how a sick girl is comforted:[4]

> . . . helped by Genius – untired comforter,
> The presence even of a stuffed Owl for her
> Can cheat the time; sending her fancy out
> To ivied castles and to moonlight skies,
> Though he can neither stir a plume, nor shout;
> Nor veil, with restless film, his staring eyes.

Clearly, this is not vintage Wordsworth; Wordsworth wrote a good deal that was not; but if we read the lines not with intent to giggle, but with intent to read, we can see that, though not great poetry, they are not silly; they are true to experience; they are about the mystery of imagination; and, apart from the rather forced and ill-chosen 'shout', they say clearly and fairly neatly what the poet wants to say. The way in which not very promising objects can, by association, trigger fantasies, and the relief of fantasy, are interesting enough aspects of psychology.

Part of the legitimate business of literary criticism is to point out faults and identify bad lines; but it should be a relatively small part; and it should, ideally, be pursued reluctantly, with regret that human endeavour rarely achieves perfection, not gleefully; gloating over the frailties of greatness rarely has a good motive.

What we want to laugh at as bad writing is sometimes not even bad; we may be missing the author's intention or applying narrow criteria; but, even when we have found a dismally bad bit of Shakespeare, Wordsworth, Tennyson or Browning (and we can find one easily enough), we should still bear in mind that this is a bit of ineptitude in the midst of achievements far beyond our own capacities.

It was not an ass, but a man of rare and original intellect and a fine poet himself (not in English) who once told me he had no use for Wordsworth, and defended this by saying that anyone who could write a poem *On Seeing a Needlecase in the Form of a Harp* could not have much merit.

Wordsworth really did write this poem – it is no. xviii in 'Poems of the Fancy' – it is oddly un-Wordsworthian and I doubt if any critic would praise it highly, but if it were abysmally fatuous it could in no way detract from *The Prelude*, the best of *The Excursion*, the great ballads and sonnets and so on. A small silliness cannot nullify a mighty wisdom; a ludicrous subject can prove no more than that one day the poet chose a ludicrous subject. The right poet might have made a fine poem about this needlecase; perhaps Philip Larkin in one way, John Betjeman in another, could do something unforgettable with it?

If we actually read Wordsworth's verses, we find that they seem to be an attempt at drawing-room verse; he is apparently aiming at a pretty facetiousness, possibly thinking of *The Rape of the Lock*? It does not come off very well, though the last line,

> 'Love *stoops* as fondly as he soars',

perhaps suggests a better poem that never came; and even in

these uneasy verses there is a perceptible striving for the right word.

'E.M.S.', who made the needlecase, was Southey's daughter Edith, an intimate friend of Wordsworth's daughter Dora, who was in bad health; if Wordsworth attempted light verse, for which he had little talent, hoping to cheer up his sick daughter and her friend, it is to his credit as good-natured effort, though this cannot validate it as literature.[5]

Riding hobby-horses is even more dangerous than hunting stuffed owls: a too exclusive devotion to one particular critical theory may handicap us in reading the primary work, even before it leads us to reject the help of other critics.

I once received a translated essay collection, *Shakespeare in the Soviet Union*,[6] studies by Soviet scholars and theatre people. It is well worth reading; it reflects much real relish and love for Shakespeare; the essayists are visibly sensitive, intelligent, and, in many respects, informed; approaches to a universally valuable author from an alien pattern of culture are interesting and may broaden *our* cultural experience. Yet when I am asked to believe in Jack Cade, as portrayed in *Henry VI*, Part 2, as a sympathetic figure with his 'revolutionary programme' (re-read scenes IV.ii, vi, vii, viii, x to see the naïve and spiteful irrationality of Cade's ideas);[7] or that 'The political significance should be the foundation stone of the whole structure of the role for any actor who wishes to show Shakespeare's real Othello';[8] and the actor's main task is 'carrying out a decisive attack on the chauvinistic instincts of those whites who have lost the sense of their own humanity';[9] or that Hamlet's dying speech to Horatio is a 'politically militant spiritual testament' that 'illumines the story of Hamlet's struggle against reaction',[10] I cannot help seeing how, though Marxism as *one of several* lamps can throw a flood of light on literature from a fresh angle, Marxism as the *only* lamp can cause some distorting shadows.

But a psycho-analytical, or a theological, or a producer's

interpretation could be just as misleadingly one-sided. Claude Roy begins a chapter of his study of Jules Supervielle:[11]

> Monomaniacs. Those who *explain* Proust by asthma. Those who explain monotheism by the desert. Those who explain the revolution by the leaders. Those who explain glory by publicity. Those who explain the stroke by the counter-stroke. Those who explain the big end of the opera-glass by the little end. Those who *explain*.

– and think they have now said it all!

The most helpful image I know for those useful, illuminating intellectual constructions that are such good servants but can become such very bad masters comes from Arthur Koestler's novel, *Arrival and Departure*,[12] in which the hero both undergoes some psycho-analysis, and wrestles with political questions; in a letter he writes:[13]

> As children we used to be given a curious kind of puzzle to play with. It was a paper with a tangle of very thin blue and red lines. If you just looked at it you couldn't make out anything. But if you covered it with a piece of transparent red tissue-paper, the red lines of the drawing disappeared and the blue lines formed a picture – it was a clown in a circus holding a hoop and a little dog jumping through it. And if you covered the same drawing with blue tissue-paper, a roaring lion appeared chasing the clown across the ring. You can do the same thing with every mortal, living or dead. You can look at him through Sonia's tissue-paper and write a biography of Napoleon in terms of his pituitary gland as has been done; the fact that he incidentally conquered Europe will appear as a mere symptom of the activities of these two tiny lobes, the size of a pea. You can explain the message of the Prophets as epileptical foam and the Sistine Madonna as the projection of an incestuous

> dream. The method is correct and the picture in itself complete. But beware of the arrogant error of believing that it is the only one. The picture you get through the blue tissue-paper will be no less true and complete. The clown and the lion are both there, interwoven in the same pattern.

Though these quotations may serve as salutary reminders, the best cure for a too exclusive pursuit of one line of approach is the same as the one for most elementary critical frailties: repeated, careful, questioning, fully attentive reading of the actual text.

Johnson insisted that 'there is always an appeal open from criticism to nature'.[14] This we should remember. When John Wain in *The Living World of Shakespeare*[15] tells us that Iago is no superhuman fiend, but that 'anybody who has been in the army, or worked in a factory, or just knocked about with his eyes open, has met Iago',[16] and enlarges upon this with plentiful reference to the text, we hear the refreshing voice of sheer common sense, in a book in which that voice is often heard. In the midst of elaborate interpretations, critical jargon, and far-fetched hypotheses, we should keep hold of our common sense. A piece of great literature may well have many layers of meaning; but it is as well to look at the top layer first; the study of character and motive may be, even should be, related to real-life experience; an interpretation that sounds perverse in its wild originality may really be perverse.

On the other hand, we must accept that there may also be an appeal open from nature to criticism. When an African student insisted that *Othello* was a lot of nonsense, because if a husband in her village thought his wife had been unfaithful he would just give her a hiding and that would be that; and if he later found she had been innocent, he would apologize and give her a present, and that would be that; no one would make such a ridiculous fuss as in the play; she was very sensibly comparing literature with her own experience. She needed, however, eventually to understand that Shakespeare imagined Othello not as a man from her village, in a society of which he knew nothing, but as a Renaissance European with a black skin and a touch of the barbarian; and that the extreme emotions in *Othello* were very true to life in Shakespeare's world and are still intelligible in Europe today.

Everyday common sense is a vital antidote to pretentiousness, disproportion, one-track minds, extreme eccentricity; but it is not a universal solvent. We also need annotations, historical perspective, knowledge of literary conventions, and so on; and we have to try to make the very difficult differentiation between geniune *common* sense and a mere narrow subjective judgment based on our limited experiences, prejudices or even neuroses.

We often make appraisals not only ungenerous, but downright erroneous, by judging an author by some imagined absolute criterion without regard for what he set out to do. There is no sense in blaming Smollett for his lack of lyricism, Sterne for his lack of plot, Tourneur for the absence of fun and good humour, Dylan Thomas for a deficiency of cogent argument; we have to enquire how far they were successful in that aspect of the art of writing that interests them. This is not to contend that, for instance, Dick Francis is as significant as Wordsworth because Francis writes thrillers as masterly of their kind as *The Prelude* is of its kind; *Nerve* grips us for two hours; *The Prelude* throws light into the depths of our existence and we can find more in it every year; but a thriller should be judged by relevant criteria, which can perfectly well be articulated; critical principles can be devised even for limericks, graffiti or greeting-card rhymes. Anything, down to boiling potatoes, can be done well or badly; but we must not demand of a grilled steak that it be also a chocolate soufflé.

We shall misjudge a literary work if we forget that it is an artefact created in a specific artistic context: if, say, our criticism of a court masque is largely, perhaps unconsciously, based on what we want of a television situation comedy; if we do not realize that *The Shepheardes Calender*, *Lycidas* and *Thyrsis*, with their shepherds who are really in quite other occupations, are related to a pastoral tradition going back to Theocritus and keeping some vitality right into the nineteenth century; if we look for Keatsian or Tennysonian sensitivity to nature in *The Owl and the Nightingale* instead of relating it to the medieval beast-epic convention with its allegorical function. The most obvious and prevalent mistake of this kind is forgetting that a drama is not really a book to read, but a play to perform; we shall miss a great deal if we do not at least try to visualize and 'audilize' it as a play, and, for close study, bear in mind also the production conditions of the epoch.

At the same time, we must not discuss the artefact as if an artefact were somehow removed from the rest of life with our human emotions and motives, muddles and anguish and subjectivities; there are no real ivory towers and we cannot appreciate literature purely with reference to genres, literary traditions, forms, techniques, the various taxonomies of specialists: every writer was a person at least as complex, fragmented and unfathomable as every reader, and even the most elegant and formal piece of literature was born out of someone's psyche with some effort and even pain.

When we consider how complex we all are, what incomprehensible Martians even our own parents or children may seem, it is almost miraculous that anyone ever understands anything anyone else says, which is what the study of literature is all about; but the glorious fragmentary intimacies, in life and in literature, do occur and enrich us.

We shall never judge perfectly; we can only try to judge intelligently and sympathetically. When we have done our best to cure ourselves of some of the temptations to massive misjudgments, rooted largely in our egotism and vanity, we have also to try to avoid the mistakes springing from our mere ignorance, the honest enough judgments of what is not there, based on the countless ways of misreading.

10

SOME WAYS OF MISREADING

> He was a Master of Arts – though of what arts I never discovered – and a Senior Wrangler. That is what was stated on the School prospectus, so it must have been true; but I could never understand it, for a less quarrelsome or contentious man you could not imagine.
>
> R. Austin Freeman, *Mr Polton Explains*[1]

> The Pardoner rode in the fashion of the latest jet set.
>
> Examinee

Probably we all sometimes misread the printed word, just as we sometimes mishear what living people are trying to say to us; the least we can hope to do is to keep our bad mistakes down to a fairly low frequency by genuine attention and careful thought.

Our misreadings of literature may range from total misapprehension of an author's purpose, to lesser mistakes from unawareness of literary conventions or failure to realize that ideas have changed, down to misunderstandings of single words.

If, for instance, we miss the irony in *A Modest Proposal for Preventing the Children of Poor People from Becoming a Burthen to their Parents or Country*, and believe Swift was actually advocating the eating of Irish babies, or read *Love and Freindship* as a girl's first attempt at writing a serious romance and not as a deliberate skit, we shall ludicrously and hopelessly misunderstand the thought and the skills of Swift or Jane Austen in these works. The literary student as obtuse as this is, we may hope, rare; but Defoe's *The Shortest Way with the Dissenters* really was misunderstood in his own day;[2] and Samuel Butler's *The Fair Haven*, an ironical defence of (Victorian) Christian orthodoxy, intended to question it drastically, was on its first appearance praised by some as helpful to unfortunate doubters.

To read Butler's *Erewhon* or Johnson's *Rasselas* as adventure stories would leave us much dissatisfied and missing most of the content; on the other hand, to seek for irony or allegory in John Buchan's *The Three Hostages*, which is meant to be a straight adventure story, would be just as inappropriate.

I will tell two stories against myself. I now marvel at the ineptitude that let me, in youth, interpret *Porphyria's Lover* as a prettily charming poem about two young people playing little fantasy games in courtship. My excuse for this truly elephantine blunder must be an education so obscurantist about sex that I feared even my interpretation of this shocking, tragic monologue would earn disapproval; I hardly dared admit knowing even about play as an element in courtship. The poem as a wonderfully concise study of sexual pathology, of extreme frustration driving an unstable suitor to murder and perverse self-justification, was quite beyond me.

My misreading of Yeats's *The Apparitions* resulted from an ignorance more specialized. I applied an everyday rationalism to his refrain:

> *Fifteen apparitions have I seen;*
> *The worst a coat upon a coat-hanger.*

and thought Yeats was seeking 'strength' against the 'mystery and 'fright' by reminding himself that most weird sights have commonplace explanations – as I have often had a momentary shock seeing some 'ghost' or 'villain' before realizing it was a trick of eyesight and nerves with a dressing-gown on the door, a coat over the banisters, and so on; I believed Yeats was saying what Shakespeare said in other words:[3]

> Or in the night, imagining some fear,
> How easy is a bush supposed a bear!

Unlike my misreading of *Porphyria's Lover*, this is not, I think, a ludicrously impossible reading of the poem read in isolation; but it is a totally wrong interpretation, putting an ironical, cheerful rationalism into a poet who was a fervent explorer of the spiritual and the occult. *The Apparitions* was based on a group of occult experiences Yeats had in 1933–4; they included a vision of an empty coat on a coat-hanger, which Yeats understood as a symbol of his own death.[4]

We may misread something from failure to understand some literary convention. The 'aside' was accepted for some three hundred years as a dramatic convention, though everyday experience soon teaches us that our relief of muttering to ourselves rebelliously, affectionately or peevishly is almost invariably overheard to our embarrassment. When Caesar, even after his wife's nightmares and several bad omens, does not hear Trebonius, told to be near him, add to his polite 'Caesar, I will:' the aside '– and so near will I be That your best friends shall wish I had been further.'[5] we should realize that the aside is a recognized convention and not read into the words evidence that Caesar is unobservant or Trebonius reckless. We are not to find a sociological statement about the helplessness of unemancipated Greek women in the inaction of a Greek Chorus when someone is being murdered behind a door; that the Chorus comments, but takes no part in the action, is a convention of Greek tragedy.

We are asked to accept, in both the Elizabethan drama and the Victorian novel, that dying or desperately ill people can talk more fluently and coherently than is medically probable. In some sense, almost all dramatic works depend on a convention that people are more articulate, keep to the point in a more disciplined way and achieve neater exchanges than people in real life. We must not, for example, infer that the savage Caliban or the loutish prince Cloten are of Shakespearian sensitivity and intelligence because they sometimes use very well chosen words.

The soliloquy is another convention; it is not true that people do not talk to themselves; many sane people often do so when alone; but probably few people talk to themselves with sophisticated arts of rhetoric and poetry like Hamlet, Richard III, Marlowe's Gaveston or Webster's Bosola. We sometimes have to consider whether a soliloquy is intended as mainly self-conscious self-revelation (Hamlet's 'O! what a rogue and peasant slave am I . . .',[6] Helena's 'O! were that all . . .'[7]); unconscious self-revelation (Morose's tirade of gloating spite, alienating any sympathy we might feel if he merely disliked noise, in *The Silent Woman*[8]); economical explaining of a bit of the plot (Edgar's 'I heard myself proclaim'd; . . .'[9]); mostly fun (Bottom's soliloquy on waking after his 'most rare vision'[10]); and so on; the possibilities are many and not always sharply differentiated. A villain's confession of villainy may be a mark of penitence (Claudius' apparently real remorse, 'O! my offence is rank . . .'[11]) or merely an explanation to the audience (Iago's 'Thus do I ever make my fool my purse . . .'[12]).

Then the dramatic monologue, favoured by (among others) Browning, Hardy, Kipling, Ezra Pound, Edwin Morgan, with varied techniques, presents other conventions; first, we must not attribute to the poet the sentiments he gives to his character.

Literature can never be all-inclusive; it is bound to make some selections and use some formalities. Conventions such as the soliloquy, the epistolary novel, allegory, pastoral elegy, are widespread. Some works create special formal patterns unique to themselves, for example *Moby Dick*; *Four*

Quartets; *The Waves*; William Golding's *The Inheritors*; but pattern of some sort is everywhere. We probably oscillate, in serious literary study, between studying the work as a representation of some aspect of life, and studying it as a thing designed; perhaps looking for 'truth' and looking for 'beauty' and in the end, perhaps coming to Keats's conclusion. But the inexperienced student will certainly fall into many misreadings if he never stops to think about traditional literary conventions, individual literary devices, or the importance of pattern in art.

To the modern reader, parts of Bunyan's *The Holy War* (1682), notably the last chapter, have an unpleasantly totalitarian flavour, apparently implying that it is right to execute people not only for wicked deeds, but for mistaken opinions. Here we must consider the literary convention: Bunyan is writing allegory, not straight fiction, and is allegorically advocating, not the extermination of people, but the elimination of impious thoughts within one's own soul.

Here, however, we need to bear in mind not only a literary convention, but a change in the climate of opinion; Bunyan was writing in an epoch when religious controversies were often brutally uncharitable. Communal prejudices change, as is illustrated by the chauvinism and extreme anti-Catholic prejudice shown in *Westward Ho!* by Charles Kingsley, eminently well-meaning and in his own day progressive.

Chaucer's Monk presents Cenobia as wedded to Odenake 'in joye and in felicitee', yet:[13]

> she never wolde assente
> By no wey, that he sholde by hir lye
> But ones, for it was hir pleyn entente
> To have a child, the world to multiplye;
> And al-so sone as that she mighte espye
> That she was nat with childe with that dede,
> Than wolde she suffre him doon his fantasye

Eft-sone, and nat but ones, out of drede.
And if she were with childe at thilke cast,
Na-more sholde he pleyen thilke game
Til fully fourty dayes weren past;
Than wolde she ones suffre him do the same.

Today most of us, including most Christians, would think Cenobia in urgent need of marriage counselling and probably psychotherapy; we need to realize that to the Monk, vowed to celibacy and in the grip of a medieval moral theology that loaded sex even within marriage with guilt and fear, Cenobia was a notably virtuous wife, 'so worshipful a creature';[14] Chaucer, impersonating the Monk, was not necessarily expressing his own attitudes; he himself wrote of love and sex with warmth and sympathy; but he could write the pathetic 'retraction' found at the end of *The Parson's Tale*.

It is very easy to gasp or snigger at something as alien as this to our current assumptions. It may be amusing to read something 'old-fashioned', or that to us seems sentimental, in a guying manner, sharing with friends our sophistication and supposed freedom from illusions. We have to go far beyond that if we are to comprehend a literature of uncongenial tone or content: we have to broaden our sympathies and develop a historical imagination, and somehow grasp that what to us is an absurd, even (as the morality of Cenobia is to me) a positively warped attitude, could once, to someone else, in a different set of conditions and beliefs, seem perfectly tenable. It is very difficult to develop even a weak and patchy historical imagination. Naturally. As in human relationships, intelligent sympathy makes heavier demands upon us than flippant contempt. We should remember, too, that while intellectual progress over the past six hundred years has been staggering, we today are not at the summit of human achievement – many of the ideas we take for granted will become obsolete.

Richard II is an inadequate king, who drowns himself in self-centred rhetoric; but when he says:[15]

> Not all the water in the rough rude sea
> Can wash the balm from an anointed king;
> The breath of worldly men cannot depose
> The deputy elected by the Lord.

we must not take this as mere self-indulgent bombast; in Shakespeare's day it was a doctrine seriously believed. The play ends with the usurper, Bolingbroke, going on a penitential pilgrimage; as Henry IV he is troubled for the rest of his life;[16] even Henry V, outwardly confident, has his deep anxieties:[17]

> Not to-day, O Lord!
> O! not to-day, think not upon the fault
> My father made in compassing the crown.

To Shakespeare's audience, Richard could be shown as an unworthy 'deputy'; but the divine right of kings was a real, living concept; to depose the anointed king was frighteningly sacrilegious; if we do not realize the power of this concept, we miss much in Shakespeare's Histories.

Jane Austen is brilliant in revealing, through dialogue and action, with grace, economy and subtlety, the distinctions between selfishness and considerateness, insensitive self-importance and courteous tact, silliness and sagacity; the real insights are permanent, but some of the detailed symptoms now demand a modicum of historical imagination. Though we are meant to see Mrs Bennet's matchmaking as vulgar, insensitive and foolish, we must also enter into a social situation in which the Bennet girls do need husbands. Today five healthy girls, at least three of them fairly intelligent, could quite properly and happily support themselves independently. (The bookworm Mary might be purposefully studying for an external degree, instead of being selfishly absorbed in reading shown as pointless?) What is wrong with Mrs Bennet is not that she wants husbands to support her daughters, in an epoch when careers are not open to them, but that she cares nothing for compatibility, mutual respect or affection. In chapter 59 Mr Bennet shows a more enlightened attitude. We have, again, to make an imaginative move into another climate of opinion to accept the overwhelming grief and shame caused by Lydia's scandal, and the idea that the best of a bad job must be to bribe the worthless Wickham to marry her, though the prognosis for such a marriage is poor.

In 1978 I encountered some students who seriously contended that a significant clue to Othello's lack of real love for his wife was his acceptance of an order to part from her on her wedding night. I thought them oddly dense, until I myself made the effort of historical imagination to realize that these students, blessedly spared direct experience of war or even conscription, were not truly conscious of things all middle-aged Britons know about war; thus common sense, uninformed, could not tell them that a general, summoned to a cabinet meeting with news of an impending invasion, would assume, however regretfully, that this was an 'all leave cancelled' situation.

Subsequent intellectual progress may cause difficulties with old literature; and imagery drawn from obsolete theories may survive the theories for hundreds of years. Any book written before Copernicus' *De Revolutionibus* appeared in 1543 assumed the Ptolemaic cosmology with a small, earth-centred universe; but many writers used the superseded Ptolemaic system, at least as a source of imagery, well into the seventeenth century. Milton – who had met Galileo – has in Book VIII of *Paradise Lost* a difficult but strangely beautiful discussion of the two systems. (We are still borrowing from the Ptolemaic cosmology today when we speak of being 'in the seventh heaven' or even 'the sun going down'.)

Marlowe's Tamburlaine, saying 'Nature that framed us of four elements',[18] is referring to the four substances then believed to be the foundation of everything: fire, air, earth and water; science has gradually identified a hundred and six elements, but the literary student still needs to know something of the old theory of four; of the four humours of early physiology and psychology; of early medicine, which provides many images in literature; of early geography, even back to a time when the Americas were unknown. Sometimes we have to imagine ourselves back into an earlier ignorance, in order to attain a full knowledge.

Not only medieval writers, but probably many writers as late as the Victorian period, had a time-scale far smaller than our own. My mother's Bible, printed not much before 1919, still gave 4004 BC as the date of the Creation, the date calculated by Henry Usher in his *Annales Veteris et Novi Testamenti* (1650–4). Recent carbon-dating methods have led scientists to place Stone Age man some 18,000 years ago and recent calculations have put back the start of our universe to as far as fifteen million years ago. We should read Marvell's *To his Coy Mistress* with reference to his time-scale, not ours.[19]

A lord and his lady in Chaucer's day enjoyed nothing like the comforts, safety and entertainment available to the dustman and his wife in a council house today. Marlowe's Dr Faustus needed Mephistophilis to fetch a dish of ripe grapes in January; the dustman's wife can get them from the supermarket. Most of us have seen many animals that to Shakespeare were only distant marvels; and can sometimes travel at speeds that even Gladstone would have found unimaginable.

We often lose something by missing some allusion. Even this triviality:[20]

> A daring young lady of Guam
> Observed, 'The Pacific's so calm

> I'll swim out for a lark.'
> She met a large shark . . .
> Let us now sing the Ninetieth Psalm.

is appreciably funnier if we know that the last line is more than just a vaguely religious reference; the Ninetieth Psalm is part of the Anglican service for the Burial of the Dead. To turn from the ridiculous to the sublime, *On the Morning of Christ's Nativity* derives part of its power from evocative allusions; they are useless if they evoke nothing or the wrong thing:

> In consecrated Earth,
> And on the holy Hearth,
> The *Lars*, and *Lemures* moan with midnight plaint.
> In Urns, and Altars round,
> A drear, and dying sound
> Affrights the *Flamins* at their service quaint.

If, guessing desperately, we think *Lars* must be some Swedish hero, or Lars Porsena of Clusium; *Lemures* are those dear little monkeyish animals from Madagascar and *Flamins* are altar-lighting ceremonies on the analogy of sit-ins, our experience of Milton will be much impaired. When we know that *Lars* (an anglicized plural) were Roman household gods worshipped at the hearth, *Lemures* were Roman ghosts, appeased at an annual festival, and *Flamins* (or *Flamens*) were Roman priests, the picture coheres and contributes to the general theme of pagan gods being superseded by Jesus:

> Our Babe to shew his Godhead true,
> Can in his swadling bands controul the damned crew.

Emily Dickinson's poem *I Dreaded that First Robin So*, about the pains of spring to a broken heart, perplexes us by an apparently senseless first line – until we learn that the American 'robin' (a red-breasted thrush) is a migrant bird, so that there is in America a 'first robin' in the spring.

Auden's *In Memory of W. B. Yeats* can stand by itself, but it gives more satisfaction to someone who knows something of Yeats's life and poetry; it gives more still to someone who has read Yeats's *Under Ben Bulben* and can see how in the final part of his elegy Auden echoes the verse-form and something of the tone and style of Yeats's poem. The last line of Auden's *Epitaph on a Tyrant*, 'And when he cried the little children died in the streets', is more powerful if we recognize a bitterly ironical inversion of the once much-quoted final sentence of J. L. Motley's *The Rise of the Dutch Republic*, borrowed from an official report on the death of William the Silent:[21]

> As long as he lived, he was the guiding-star of a brave nation, and when he died the little children cried in the streets.

Giovanni's 'Suppose me one of Homer's frogs, my lord, Tossing my bulrush thus' in Webster's *The White Devil*[22] is a very suitable allusion in the mouth of a lively boy of the period; his education would be classical, and the comic *Batrachomyomachia* ('Battle of Frogs and Mice') to which he refers was then thought to be by Homer. And so on, and so on; the more full is our knowledge, the greater is our enjoyment.

We can misunderstand simply because we do not know the meaning of a word, like Shelley, who, hearing the name,

> thought
> This Aziola was some tedious woman,

until

> Mary saw my soul,
> And laughed, and said, 'Disquiet yourself not;
> 'Tis nothing but a little downy owl.'[23]

or the admirable journalist, who confessed that as a boy reading about Lady Macbeth's 'poor cat i' the adage'[24] he thought an adage must be a sort of Scottish fish-tank.[25]

Least dangerous are words so rare or so old that we at once know we do not know them (my italics):

> . . . rousing its priest, treacherously promising *vaticination* . . .
>
> Charlotte Brontë, *Villette*[26]

> . . . an indescribable consciousness of Arabella's midnight *contiguity* . . .
>
> Thomas Hardy, *Jude the Obscure*[27]

> The *golk*, the *gormaw* and the *gled* . . .
>
> William Dunbar, *The Fenyeit Freir of Tungland*[28]

> Since such loves naturall *lation* is, may still
> My love descend, and journey downe the hill, . . .
>
> John Donne, *The Autumnal*[29]

Vaticination and *contiguity* we can find in any good one-volume dictionary if they do not happen to be in our working vocabulary. In the ordinary dictionary we may find *gled* ('kite', the bird) but for *golk* and *gormaw* ('cuckoo' and 'cormorant') we shall need a specialist glossary such as W. M. Mackenzie's edition of Dunbar provides, or access to the *OED*. For Donne's *lation* we need a detailed annotation and must turn to Helen Gardner's edition of *The Elegies and the Songs and Sonnets*.[30]

When we know that we do not know, we can look up the word: we fall into traps mostly when we think we know. In Donne's *The Sunne Rising*, we meet 'Love, all alike, no season knowes, nor *clyme* . . .'. If we assume *clyme* is an old form of *climate*, our reading will make sense; but it will be wrong; *clyme*, now always spelt *clime*, is related to *climate* etymologically, but means *region*. (Indeed, Shakespeare uses *climate* in the sense *region* in the line, 'Unto the *climate* that they point upon.'[31])

Milton's serpent, going to tempt Eve, has[32]

> his Head
> Crested aloft, and *Carbuncle* his Eyes;

Though evil, he is beautiful; carbuncle here does not mean, as it usually does today, a nasty septic swelling, but a red precious stone. *Dangerous* in Chaucer and other medieval writers means something approaching 'hard to please', usually in a courtship situation; students of medieval literature need to know something of the special concept of 'danger' in the courtly love conventions. *Gentle* shifted towards its modern meanings of 'mild, not severe, tender' only at about the end of the sixteenth century; earlier, it is likely to mean 'well-born, noble, spirited as a nobleman should be' and the like: Miranda's 'Make not too rash a trial of him, for He's gentle, and not fearful'[33] does not mean 'Do not bully him unthinkingly; he is mild and polite, not a coward' but something more like 'Do not risk bullying him; he is of high birth and not a coward', so 'he is noble and spirited and will resist'.

Candour today means 'frankness', but earlier could mean 'brilliant whiteness', 'innocence', 'purity' or 'impartiality'. *Honest* could once mean 'chaste'; *ecstasy* could mean 'extreme distress' or even 'fainting fit'; *fulsome*, now meaning 'excessive', usually with reference to flattery or displays of affection, could earlier mean 'loathsome', 'foul', 'gross', 'lustful' or even 'plump'. And only recently I learned that in Cornwall *figs* used to mean 'raisins'![34]

We must beware of real ambiguities: *moped* may mean 'was depressed' or 'in a depressed state', or 'motor-assisted bicycle'; *denier* may mean 'someone who denies', 'a denial', an old coin, or a unit of fineness in thread, especially in stockings; *parts* may mean 'portions', 'divides, separates', 'departs', 'qualities', 'talents', 'bodily organs'; and so on. Then there are ambiguities intended by writers, as in puns:

> And see! thy very very bands
> Are bound to thee, to bind such hands.
>
> Richard Lovelace, *A Guiltless Lady Imprisoned; After, Penanced*

and more subtle multiple meanings:

> When we have run our Passions heat,
> Love hither makes his best retreat.
>
> Andrew Marvell, *The Garden*

We can sometimes fall victim to our own irrelevant associations; a cat-lover and ignorant of motor-cycling, I saw a headline, 'Clear road for Manx G.P.'[35] and momentarily pictured a tail-less cat with a doctor's bag walking to a patient. The article was about the Grand Prix in the Isle of Man. I have seen a student trapped into asserting that Chaucer emphasizes that his Knight is fully heterosexual: 'His hors were gode, but he was not gay.'[36] *Gay* in the sense of 'homosexual' is not recorded before 1935.[37]

Finally, we need sometimes to be sure we are seeing the word that is there. As a child not yet knowing *hospitable*, I read the title of an anecdote as *The Hospital Cat* and was puzzled to find not so much as a nurse in it; recently I saw *indigenisation*, in an article about Nigeria, first as *indigestion*. Examinees have informed me that Satan was 'Vaulting aloud, but racked with deep despair' (*Vaunting*); that Lear intended to keep 'The name, and all the addiction of a king' (*addition*, i.e. 'honours', 'titles', 'ceremonial observances'); that Goneril declared herself 'worth the thistle' (*whistle*) and even that the boy Wordsworth's boat was 'an elfin pinnacle' (*pinnace*).

We must not leap into amateur textual emendation too hastily; during a hasty reading through *The Mysteries of Udolpho* I met 'the *izard* among the rocks' and decided, with all the self-confidence of semi-attention, that it was a misprint for *lizard*, until some pages later I met a hunter chasing 'the izard, or wolf' and looked up *izard* to find it was a Pyrenean antelope!

The great, good, reticent scholar, J. C. Maxwell, once told a friend that the only epitaph he would like to have on his tomb would be 'He got it right'.[38] He did; and it is a mark not only of the great scholar, but even of the true student, to try to get it right. The beginning of getting it right is the painful awareness of how likely we are to get it wrong.

11

FIGS, DATES AND REASONS

'Don't imperrupt!' he said as we came in. 'I'm counting the Pigs in the field!'

'How many are there?' I enquired.

'About a thousand and four,' said Bruno.

'You mean "about a thousand",' Sylvie corrected him. 'There's no good saying, "and four": you can't be sure.'

'And you're as wrong as ever!' Bruno exclaimed triumphantly. 'It's just the *four* I *can* be sure about; 'cause they're here, grubbling under the window! It's the *thousand* I isn't pruffickly sure about!'

Lewis Carroll, *Sylvie and Bruno Concluded*[1]

However sensitive our responses, however ripe our critical judgment, however profound our insights, we need also some equipment of facts; even of facts dry and empty enough in themselves to satisfy Mr Gradgrind. Checking points of fact not in themselves interesting can save us from mistakes so silly as to be humiliating, or from brilliantly original hypotheses that happen to be impossible. We need to know that the Samuel Butler of *Hudibras* is not the Samuel Butler who wrote *Erewhon*; that Petrarch and Plutarch, though both had a substantial influence on our sixteenth-century literature, are not alternative versions of the same name; that it was not Hugh Walpole, but Horace Walpole, who wrote *The Castle of Otranto*; that Ben the dramatist spelt his name

Jonson and Sam the lexicographer spelt his *Johnson*; that free verse and blank verse are far from being the same; that until 1930 the name Pluto cannot refer to the planet; that in Shakespeare's day historical dramas were performed in contemporary, not historical, costume; that nothing was *printed* in English till 1475, so no literature could be widely distributed or have a 'public' in any modern sense.

People who are interested principally in verbal communications often dislike working with numbers. I know I do. However, mathematicians are not excused from answering letters or talking to friends, and the highly literate are handicapped if they insist on being almost innumerate.

We do not need to know many dates by heart, but we do need to take some notice of dates and to check them whenever dealing with a matter to which chronology may be relevant.

If, for instance, we read F. T. Prince's fine poem, *Soldiers Bathing*, in an anthology, we may wish to know which war gave rise to it. There is no internal evidence: no reference, say, to the causes of the war, to specific weapons or a specific location; indeed, references to Italian pictures imply that in some sense all wars are one and all tortured soldiers are related to Christ crucified. The technique makes it almost certain that the poem is of the present century; though it has formal structure and the flexible couplets rhyme firmly, yet the variability of the couplets is of a kind unlikely in previous centuries. When we learn that F. T. Prince's first volume appeared in 1954, it seems likely that the poem is from the Second World War. If we can then find Prince's birth date, which is 1912, it is clear that he did not serve in the First World War. These data do not absolutely prove that the poem relates to the Second World War; any poet may write imaginatively about a war in which he took no part; but the data bring us nearer to a sensible hypothesis.

If we do not find as much originality in *The Voyage Out* as in *Mrs Dalloway*, the fact that the former was Virginia Woolf's first novel, published in 1915, and the latter, published in 1925, was a novel of her prime, is relevant to critical discussion. Shaw's *Buoyant Billions* is very disappointing Shaw, though not without wit, invention and provocativeness; but if we look up Shaw's birth date, 1856, and subtract it from the date of *Buoyant Billions*, 1947, we realize that it is the work of a man of *ninety-one*, so the amazing thing is not the deterioration, but the merit!

We sample Crabbe, an author more worth reading than read: on the most casual inspection, we see that he almost always uses heroic couplets; after a few minutes' reading we begin to think of Pope:[2]

> This love of life, which in our nature rules,
> To vile imposture makes us dupes and tools;
> Then pain compels thy impatient soul to seize
> On promised hopes of instantaneous ease;

Did Pope refine the techniques of Crabbe, or did Crabbe learn some techniques from Pope? We need go no further for our answer than a reference book that tells us Pope died in 1744 and Crabbe was born in 1754: that settles it.

I was once inclined to think of Dryden and Pope as part of the same movement; but a glance at dates tells us that Dryden died when Pope was a boy of twelve. We may think of The Romantic Movement as a school of poets working together. Biographies will make clear that this is not so; only Wordsworth-Coleridge-and-Southey and Byron-and-Shelley had much to do with one another; Wordsworth had one or two brief meetings with Keats and never met Shelley; Wordsworth, eighteen years older than Byron, twenty-two years older than Shelley and twenty-five years older than Keats, was a man of a different generation; he also survived Byron by twenty-six, Shelley by twenty-eight and Keats by twenty-nine years, living into a considerably changed intellectual climate.

Checking dates may sometimes improve our understanding of lesser details. Byron, in *English Bards and Scotch Reviewers* (1809) was highly offensive about Wordsworth and Coleridge[3] – unfair, and not even accurate; but this does not mean Byron was wholly and permanently incapable of appreciating what was best in them. Some checking in reference books tells us that he wrote the insulting lines when only twenty-one; so some immaturity of judgment was forgivable; that he could not have seen *The Prelude* or *The Excursion*, *Christabel* or *Kubla Khan*. If we then look at some of the works Byron *had* seen, we may perhaps grant that, though he completely misses what is subtle and noble in *The Idiot Boy*, it is not the easiest poem for a rather cynical youth to appreciate; and that when he jeers at Coleridge who 'takes a pixy for a muse' and 'soars to elegise an ass', he is not alluding to Coleridge's best work: *Songs of the Pixies* comes near to justifying Byron's 'tumid stanza', and *To a Young Ass*, however creditable in its compassionate sentiments, is not much better.

We may also learn from a biography of Byron that he later became friendly with Coleridge, apologized for his satire as 'pert, and petulant, and shallow enough'[4] and sought to help him; dined with Wordsworth[5] and developed some degree of respect for him, though he never found him really congenial;[6] and as a final judgment on *English Bards and Scotch Reviewers* wrote, 'The greater part of this satire I most sincerely wish had never been written. . . .'[7]

Biographical detail may sometimes modify our more general critical judgments. If we read Shelley's *Stanzas Written in Dejection*, near Naples, we may be tempted either to identify ourselves with the tone of despair, in a luxuriating youthful depression, or to be impatient with the self-pity and think Shelley is luxuriating in youthful depression. If, however, knowing the poem was written in December 1818, we consult a biography of Shelley, we find that when he wrote those stanzas: he was suffering severely with, probably, pulmonary tuberculosis, and undergoing painful treatment;[8] the Lord Chancellor had penalized his unconventional lifestyle and opinions by depriving him of access to his two children by his previous marriage;[9] he had been savaged by the *Quarterly Review* and some really ugly calumnies were circulating;[10] at the same time he was trying to cope sympathetically with Claire Clairmont's troubles and other domestic problems;[11] and his baby daughter by his wife Mary had died in September;[12] a load of real pain and grief sufficient to overwhelm someone less sensitive than Shelley. Now we realize that the lamentations in the poem are fully reasonable; and it is evidence of Shelley's courage and dedication that he could express his misery in such controlled stanzas and with the beautiful pictures of the Mediterranean scene.

Biographical detail is not a substitute for literary criticism: a poem or novel or play, or even an autobiography, has ultimately to stand or fall on its own merits as a work of art, an artefact in isolation; but biographical or historical knowledge may help us not only to judge a work of art more sympathetically in relation to the author, but to make more sense of it, which clearly is relevant to real critical judgment. Often the factual detail we need, though in one sense extraneous to the work, is in another sense not only legitimately relevant, but vital; the topic was common property to the public for which the work was written; the author could take it for granted that he did not need to give preliminary explanations; but, anything up to twelve hundred years afterwards, *we* may need explanations. We need to be told that *The Boke of the Duchesse* (1369) was an elegy for a real person, Blanche, first wife of John of Gaunt, Chaucer's patron; Chaucer knew them both, and the poem is more touching when we realize the delicacy and gentleness with which Chaucer seeks to convey personal sympathy, without intrusive presumption, through a veil of medieval, dream-convention and other devices.

But today's undergraduate may already also need some information about the Spanish Civil War and its significance for many non-Spaniards, in the 1930s, to understand Auden's *Spain* (1937) or C. Day Lewis's *The Nabara* (1938).

Literary studies sometimes include what might be called gossip round a literary work rather than contributions to critical understanding; gossip, though it may humanize a writer for us or make a book seem more interesting, is not strictly relevant; explanation is. The story of how John Stuart Mill's servant destroyed the first manuscript of Carlyle's *The French Revolution* is moving, in the picture of an enormous labour all to do again, in the generosity and magnanimity recorded on both sides, above all in the scholarly heroism and stoicism of Carlyle's rewriting; we may well be the better for such examples; but that story does not make that marvellous work any greater as a *literary* achievement, though it reveals it as a greater personal victory of an individual spirit. On the other hand, Dr H. Ben-Israel's article, 'Carlyle and the French Revolution',[13] does contribute to our actual appraisal of the book, in showing how far it is based on careful scholarship with a real search for historical truth; if Carlyle's unique and haunting prose poem were largely untrue, it would be less great as a literary work.

Certain advanced and minute literary researches now make use of statistical methods and even of computers; the first-degree student need not consider such techniques; but a few simple statistics, such as can be compiled by anyone who can count, are sometimes useful for confirming or confuting our first opinions and for presenting evidence.

Some of William Blake's poems are well known; how much of the corpus of his poetry has really become a part of the national culture? Let us look at Blake's Complete Works and ask ourselves what works are fairly well known? We shall, I think, come up with: perhaps four of the *Poetical Sketches*; all of *Songs of Innocence and Experience* and about half of the *Miscellaneous Poems and Fragments*, which include *Auguries of Innocence*; perhaps *The Everlasting Gospel*; a few lines from *The Marriage of Heaven and Hell*; Blake's best-known poem, 'And did those feet in ancient time . . .', but not *Milton*, to which it is prefatory; possibly 'I saw a Monk of Charlemaine' in *Jerusalem*; possibly *The Gates of Paradise*; a few epigrams; certainly no more. It takes only a few minutes to transpose this into figures: the Nonesuch complete *Poetry and Prose* has 926 pages of actual text; the works listed above add up to about 77 pages, or about 8 per cent. Even if someone reasonably objects that the Nonesuch edition includes a lot of marginalia and letters that are not organized works of literature, and we take the true 'Works' as being 700 pages, the well-known works form no more than 11 per cent of the Blake corpus. We may draw various inferences from this.

We are told, or feel inclined to say, that Thomas Hardy is very fond of rather rare words, and of modifying commoner words with affixes or grammatical endings such as are understood at once, but are unconventional in their application. We do not want to make wild generalizations; have we just noticed a few words and deduced a thicker sprinkling than is there?

Now suppose we take *Wessex Poems*, a small volume, 228 pages in the well-known 'Pocket Hardy' edition from Macmillan; and quickly run up two lists. This took me minutes, not hours; I may have missed something, and obviously individuals will differ slightly as to what is a rare word. But here are my lists: *aethered*, *aureate*, *blee*, *caddle*, *capple*, *cicatrize*, *coll*, *dorp*, *durn*, *effulgent*, *engrailed*, *fane*, *flambeaux*, *gallied*, *ghast*, *graveacre*, *grinterns*, *grizzel*, *hodiernal*, *huddied*, *illuded*, *illumes*, *impercipient*, *ingresses*, *intermissive*, *intervolve*, *jee*, *joyance*, *junctive*, *knap*, *leazes*, *lewth*, *linhay*, *lumpered*, *lynchet*, *mew* (a place), *mixen*, *moils*, *nescience*, *nimb*, *ostent*, *poussettes*, *prevision*, *rayed* (dressed), *sempiternal*, *shrammed*, *slats*, *snocks*, *stillicide*, *subtrude*, *supernal*, *surcease*, *tallet*, *tardle*, *thirtover*, *touse*, *trine*, *vlankers*, *yestereve*: fifty-nine rare words; then *acheful*, *chancefulness*, *childing*, *disennoble*, *embowment*, *enarch*, *forefelt*, *forthcome*, *gagèd*, *hazardry*, *indwell*, *inscrolls*, *inurned*, *irised*, *lippings*, *lotted*, *maledict*, *miscompose*, *nighed*, *outflee*, *pilgrimed*, *pupilage*, *ranksmen*, *sheened*, *showance*, *sicklied*, *smugger*, *tarriance*, *tendance*, *unbe*, *unchosen*, *undulled*, *unknows*, *untrumped*, *upcloses*, *updrave*, *upglasses*, *upspoke*, *vigiling*, *wayless*: forty-one further words in which Hardy modifies commoner words with the freedom of Shakespeare, e. e. cummings or Esperanto. My total, then, is an unexpectedly neat hundred.

We might have said, on general impression, 'Oh, Hardy has a curious word on every other page!' This has such a tone of doorstep gossip that it sounds more slovenly than it is; if we can say that we have checked that in 228 pages, forty-four are illustrations or blanks, so that there are 184 pages of text, and we have our hundred or so words, we can conscientiously and confidently say, 'Hardy uses, on average, a curious word rather more often than on every other page, in *Wessex Poems*.'

Similarly, we have gathered a vague notion that Keats likes to invent expressive hyphenated words. It took me only fifty minutes to run through *Endymion* and list 181 such words (omitting obvious everyday words such as *elbow-deep* or *blear-eyed*). Keats's words include such happy concisenesses as *men-slugs*, *near-dwellers*, *merry-winged*, *half-graspable*, *thunder-gloomings* and *eye-guess*. If I were about to write an essay on *Endymion*, I could now treat of Keats's use of compounds with more than a half-graspable eye-guess.

Figures, dates and tabulations may not contribute very much to our literary studies; but they are some protection against sweeping statements, woolly wanderings, anachronisms and nonsense.

12

BACKGROUND KNOWLEDGE

> That you have a saturation point of interest tells us nothing of the interests that absolutely are.
>
> William James, *Human Immortality*[1]

There is always something more to be learned about any subject that is worth studying; we are never totally equipped to understand and appreciate the whole range of English literature, any more than we can ever be adequate in all personal relationships. The right inference from this commonplace of experience is not a despairing laziness, but some sensible humility.

Any serious student must, however, have certain stocks of background knowledge if he is not to miss the point more often than he sees it. At best it will be scrappy, but here scraps are better than nothing. The student of literature almost always has a struggle to pick up some of the necessary background knowledge concurrently with his actual literary studies, and has not much time to learn things he should, ideally, have known before embarking upon the course. Part of the pre-university summer holiday of the school-leaver might well be spent, if possible, on assembling some basic background kit, minimal though that will be. The busy student may well feel that to have adequate background knowledge he would need to take four or five degree courses – he would; but he can slant some recreational reading in suitable directions.

A reminder that books and lectures are not the only sources of cultural information may not be redundant. Eyes that ache from reading can rest while a radio programme gives some useful knowledge; maybe simplified and popularized, but probably sound and helpful. There are often relevant and agreeable television programmes. And all round the English Literature student there are other students reading other subjects. The student in your year reading Classics is no more a classical scholar than you are yet an English scholar; but he can tell you who Atalanta was; the History student probably knows more about laws and life in Shakespeare's day than you do; a wrestle with William Empson's very demanding poems may be helped a little if an ecologist friend annotates 'Daily brings rabbits to a new Australia'[2] or a budding psychologist can at least place 'Piaget's babies' who 'spoke not to Piaget but to themselves'.[3] Friendships extending into other faculties not only give us some happiness and hope of giving happiness, and broaden our minds, but provide us with reference books on legs. Most people quite enjoy a chance to display their superior knowledge.

The following inventory of equipment for exploring English Literature may look formidable; but it is far from comprehensive; that, no such inventory could ever be.

1 Anyone who means to make any analytical approach to literature must have an adequate vocabulary of ordinary critical terms. There is certainly no merit in a lot of pretentious jargon; but it is hardly possible to describe precisely a poet's handling of a verse form unless we know how to scan and command some prosodic terminology, or to discuss his imagery without knowing the names for at least the commoner figures of speech. We can manage without, say, *epanadiplosis*, but hardly without *metaphor* or *alliteration* or *personification*. We must know something of literary forms and conventions, terms such as *epic*, *pastoral*, *allegory*, *satire*, *lyric*, *madrigal*, *fable*, *parody*, *heroic drama*, *chronicle play*, *masque*,

picaresque novel, *Gothic novel*, *naturalism*. We need to know what we mean by such terms as *objective*, *sentimental*, *farcical*, *rhetorical*, *symbolism*, *Renaissance*, *Georgian* (and to check that we do mean what previous users have meant; sloppy usage may produce a smokescreen rather than a searchlight . . .). We can get a long way without the special terms of particular critics, for instance Matthew Arnold's 'Hebraism and Hellenism' or T. S. Eliot's 'objective correlative', unless we are discussing Arnold's or Eliot's criticism; but we must have a basic critical vocabulary.

2 English literature is a part of English history, and inevitably reflects it; so we need some English history. There are aspects of Milton or Marvell we cannot appreciate without some idea of the English Civil War, the Commonwealth and the Restoration. We also need some European history; *The Prelude* is not fully intelligible without some knowledge of the French Revolution and the resultant wars. And if the syllabus includes some American literature, the rudiments of American history will also be needed. It is probably true to say that as communications grow faster and easier we need a wider field of history and politics for literary studies: a twentieth-century writer may be moved to write of events almost anywhere, as, for instance, when Louis MacNeice's *Didymus*[4] treats of Thomas the Apostle in the setting of a modern man's experience of India, or D. J. Enright writes of 'The Burning of the Pipes' in Bangkok in 1959.[5]

We also need as much Ancient History as we can pick up; for instance, some of Plutarch's *Lives* for Shakespeare's Roman plays. Allusions to Roman and Greek history crop up throughout English literature, because for centuries they were part of an English boy's education. When Robert Burton wrote, 'Socrates his *cicuta*, Lucretias dagger, Timons halter are yet to be had; Catoes knife, and Neroes sword are left behind them, as so many fatal engines, bequeathed to posterity . . .'[6] any educated reader of his day would immediately understand all the allusions; so too would an educated contemporary of Byron catch the right evocations from:[7]

> I roam
> By Thrasimene's lake, in the defiles
> Fatal to Roman rashness . . .

3 We need a great deal, not only of Ancient History, but of Greek and Roman background in general. The classical education that remained dominant for centuries in Europe had its obvious disadvantages; notably, it had nothing to offer the boy whose talents were not linguistic, and in centuries in which severe corporal punishment was normal practice must often have involved torture as well as boredom; but it also had its merits. Both Greece and Rome offered rich, varied, sophisticated literatures and complex, mature languages; the classically educated had a common body of stories, allusions and verbal techniques well worth having; British civilization thus kept roots in the great Mediterranean cultures as well as in the North; the classics may well have been far more broadening than narrowing in their total effect. The influence extended to American literature, though it was not quite so pervasive. This classical influence is still a strong one; we see it, for instance, in the wickedly funny burlesque of Oedipus Rex in John Barth's *Giles Goat-Boy*,[8] Ezra Pound's *Homage to Sextus Propertius* or allusions in the poems of Peter Levi. At the time of writing the youngest significant poet whose published work I know, Andrew Harvey, who certainly does not lack originality, takes off from Horace for several of his best poems.[9]

Yet today many readers miss the complex evocativeness of Housman's:[10]

> By Sestos town, in Hero's tower,
> On Hero's heart Leander lies;

The signal torch has burned its hour
And sputters as it dies.

and Sir John Davies's rhetorical question,

Where lives the man that never yet did hear
Of chaste Penelope, Ulysses' queen?[11]

might now be answered, disappointingly, 'In lots of places, even this university'.

The student who enters the university with no Latin or Greek cannot be expected to learn even one difficult dead language on top of everything else; but if he has also arrived with no knowledge of classical references, he is in desperate need of instruction. The mythology is essential equipment: the names of the gods and their associates in both languages (Zeus/Jupiter/Jove; Pallas Athene/Minerva; Poseidon/Neptune and so on) and the chief powers attributed to them; the stories about them; the legends about the constellations (e.g. Orion, Cassiopeia); the matter of the *Iliad*, *Odyssey* and *Aeneid*, and the best-known Greek tragedies; something of Theocritus and the *Eclogues* and *Georgics* of Virgil, for the pastoral tradition; Horace's *Art of Poetry*, relevant both to the general principles many writers had seen, and to Pope's *Essay on Criticism* or Byron's *Hints from Horace*; something of Ovid, Catullus, Juvenal (the model for much verse satire), Lucretius; this list is an example and a suggestion for some priorities, not even a minimum.

What is the non-classicist student to do? He can at least acquire a classical dictionary and look up allusions as often as he can; and if he can manage to read through the classical dictionary bit by bit in odd moments it will be some help. This sounds rather dreary, but, after, all, Lemprière's *Classical Dictionary* inspired Keats. The classical dictionary will identify names and furnish the essentials of many stories; but it cannot show us great literary traditions or bring the great legends to vivid life. The best classical education without Greek or Latin is to read as many translations as possible. Sometimes we may manage to read translations by writers who themselves have a place in our literature: we can see what looking into Chapman's Homer does for us, or try Pope's Homer; Marlowe's version of Ovid's *Elegies*; Golding's of Ovid's *Metamorphoses*; North's Plutarch, which gave so much to Shakespeare; Dryden's Juvenal; Shelley's version of Plato's *Banquet* and of the 'Homeric Hymns' (no longer attributed to Homer, but important sources of mythology); C. Day Lewis's translations of Virgil; MacNeice's of the *Agamemnon* of Aeschylus – or try the Browning version; from such reading we at the same time pick up some classical background and learn more about the British authors and the influences acting upon them; thus we achieve what in this classical mood I will call 'duplex unilapidarian avicide'. The great literary translations are open to criticism; but it may well be impossible to make a translation that is not, and for present purposes we want the stuff of a cultural heritage rather than the fidelity of a crib. Then there are other translations: the Loeb Classics, the many translations in the Everyman and World's Classics series and now the more modern translations easily accessible in the Penguin series.

4 It is impossible to understand English literature without some knowledge of the Bible, Christian belief and tradition and some rudiments of Church history. Until very recently, a standard education in Britain included an enormous dose of Bible knowledge. It may often have been so badly presented as to be counter-productive; it could (as I know from experience) lay an excessive load of guilt and fear on a young soul; but at least it transmitted important items of Western culture. Religious education today is often broader, takes into account our multifarious society, allows a greater spirit of enquiry and sincerer involvement in ethical problems; well handled, this is in many ways admirable; but

as academic equipment for literary studies it is deficient.

In studying English literature we are looking back into a world in which:

a for nearly nine centuries (Caedmon's writings before 670 to the breakaway of the Anglican church in 1534) virtually the only possible ideological assumptions were some form of what we would now call 'Catholicism', though medieval 'Catholicism' was far more austere, superstitious and literalist than anything an English priest would wish us to accept today;

b then from about 1534 to 1688 religious controversies were among the leading issues of the day; played an essential part in the Civil War and the ruin of the Stuarts; and were pursued with a fanaticism and cruelty now more usually devoted in some countries to political or racial disputes. Though after 1688 these passions cooled and we soon find the well-known eighteenth-century distrust of 'enthusiasm', the Test Act of 1673, which prevented Catholics and Dissenters from holding office, was repealed only in 1828–9, and the University Tests Act only in 1871. The Victorian era was still full of religious controversies conducted in a more humane manner, with agonies of 'Doubt' when Darwinism and the application of developing scholarly techniques to the Bible came into conflict with fundamentalism; and on the other hand much emotion about 'Romanizing' in relation to the Tractarian Movement. We have only to read *In Memoriam* or *Robert Elsmere* to catch something of this. In 1848 Arthur Hugh Clough, a man of integrity, sacrificed an Oxford fellowship because he could no longer subscribe to the Thirty-Nine articles, still obligatory.

Also

c The Authorized Version translation of the Bible, of 1611, was one of the major influences on English style for some two hundred and fifty years. (The next translation appeared 1880–5, but has never been so popular.)

The above reminders are gross over-simplifications; the student who has already muttered, 'What about the Lollards?' or 'What about Marlowe, Bernard de Mandeville, Thomas Paine?' is right, and does not need the reminders; but many do, and have to start from scratch.

Indispensable minimal knowledge for the student of literature includes: Genesis, Exodus, Samuel I and II, Kings I and II, Psalms, Proverbs, Isaiah; the Gospels and Acts and Revelation; and at least one reading of the entire Bible is highly advisable; all in the Authorized Version, preferably.

Some elementary idea of theological doctrines and concepts is needed, notably some idea of differences between Catholic and Protestant, and what the Reformation was all about. It was *not* merely that Henry VIII wanted to marry Anne Boleyn!

The student of medieval literature needs to have some notion of the swarm of saints (many of them now repudiated by the Vatican) and mass of traditional beliefs not found in the Bible. *The Golden Legend*, if accessible, is a medieval manual of such lore: it is fascinating reading, with some charming anecdotes and spiritual insights buried in an amazing farrago of grotesque and sometimes probably pathological fantasies. Donald Attwater's *Penguin Dictionary of Saints* (1965), also fascinating, is more reliable as information but does not so much illustrate medieval background. For the study of the literature of any period we need some notion of the religious controversies and current ideas of the period.

Some idea of Christian symbols is also helpful; we shall often meet references to such things as the keys of Saint Peter, the pelican as Christ, the vine as Christ, the Holy Grail, the Lamb of God, the Scarlet Woman; the shepherd image, important in pastoral; the images of blood, water, bread and wine, light and darkness. It is useful to know of the Seven

Deadly Sins, the four cardinal virtues and the three theological virtues. We need some idea of such medieval embroideries upon Christianity as the nine orders of angels, a menagerie of classified and named demons; of monasticism and pilgrimages and of beliefs about witches. We shall find references to odd bits of popular 'Christian' lore such as Ophelia's 'They say the owl was a baker's daughter'[12] or the tradition that Judas had red hair.

Incidentally, we often need to look out for conflations of Christian and classical allusion and imagery; 'Pan' may refer to the Christian God, as in the fifth Eclogue of *The Shepheardes Calender*; Milton invokes the Muse Urania to help him 'justify the ways of God to man'; Donne pleads to God:[13]

> Oh! of thine only worthy blood,
> And my teares, make a heavenly Lethean flood . . .

Naturally, when two immensely powerful cultural traditions both provide images and associations, there will sometimes be mixing.

5 The student needs some background of medieval science, such as Ptolemaic astronomy, astrology, the theory of humours, ideas on dreams and visions, early medicine, alchemy; something of medieval society, such as the feudal system and the courtly love tradition. These things are not needed for the study of medieval literature only. Writers do not confine their subject-matter to their own epochs; the strangeness of the past nourishes our imaginations, and the history play was one of the major forms of sixteenth-century drama – still not extinct; the historical novel has been an important genre of the novel ever since *Waverley*; many of Browning's poems are set in previous centuries. Apart from imaginative presentations of historical matter, there is always a national stock of images that is not exclusive to an epoch, but is cumulative through the epochs. Robert Graves brings

two worlds together, significantly, when he portrays in 'Grotesques' a psycho-analyst producing a sooterkin from his pocket.[14] Auden and John Heath-Stubbs both wrote poems about Saint Cecilia. James Kirkup, in his witty but awed poem, 'Tea in a Space-Ship',[15] uses on so notably topical a theme the image of a *bilboquet*, a medieval toy (seen in Olivier's film of *Henry V*).

6 There is no need for literary students to learn to milk cows; but it is a help to have looked at a farm, taken some country walks and had some slight personal experience of wild flowers, wild birds, crops, beehives, horses; to have truly taken it in that earlier writers saw ducks in ponds not in deep-freezes, and were familiar with all the stages of bread from sowing to baking; that George Eliot had heard people 'discoursing . . . about the limited amount of milk that was to be spared for butter and cheese so long as the calves were not all weaned, and a large quantity but inferior quality of milk yielded by the shorthorn . . .'[16] in a world in which the horse provided the fastest known transport. Students who have grown up in cities can lose some of the vividness of narrative or the point of imagery by their ignorance of the agricultural life that for centuries was the background to most British experience.

First-hand experience can be reinforced, or, if real muddy lanes, cows, stiles and live Chauntecleer and Pertelote are truly inaccessible, imperfectly replaced, by books, especially books with good pictures, and television. Even something as simple as Wilfrid Gibson's[17]

> Up against the sky,
> Beyond the spinney and the stream,
> With easy stride and steady eye
> He saw his father drive his team,
> Turning the red marl gleaming wet
> Into long furrows clean and true:

is better realized by the reader who has *seen* ploughing, even with a tractor instead of horses.

A little rural knowledge contributes not only to the impact and thus to the enjoyment of the work; it is also relevant to criticism. Our appreciation of the precise wording of accurate, loving observation of nature by, say, John Clare, Keats, Tennyson, Hopkins, is fuller and more sincere if we can check it against our own observation. And we can more reliably distinguish between realistic treatment of rustic life, and idealizations found in pastoral and idyll. (Both may be found in Shakespeare, often with rapid shifts from one to the other; here is an interesting exercise!)

The student from abroad, still more the student studying outside Britain, will be specially remote from this background, as Louis MacNeice amusingly warns:[18]

> Wee sleekit courin' timorous warthog!
> Tirra lirra by Kabul River!
> The elmtree bole is in tiny leaf but
> Not for long because of the termites.

I heard of a lecturer who, taking a walk, saw an Indian pupil, leaning on a gate, gazing into a meadow, who responded to his greeting with an eager, 'Oh, sir! now I really understand Wordsworth's beautiful lines for the first time! look, sir! "My heart with pleasure fills, And dances with the daffodils."' Looking into the moist and luminous brown eyes, the lecturer found it hard to have to explain that the jocund company was a meadow full of *dandelions*.

Let us not forget, however, that though dandelions cannot flutter and dance with the unique beauty of daffodils, a student who has been moved to tears by a field full of golden dandelions has an incalculably better understanding of '*I Wandered lonely as a cloud* . . .' than a student who has never thrilled to the beauty of wild flowers.

7 A rudimentary knowledge of the story of exploration is

useful: we need to realize that the *Mayflower* sailed towards the first British settlement in America four years after Shakespeare's death; that Australia and New Zealand were totally unknown to us until Tasman's voyages in the late seventeenth century, real knowledge beginning only in the late eighteenth century with Cook; that pioneer explorations in Africa were still going on in the Victorian epoch and that much exploration of South America is of our own century. Old maps can be very enlightening. Donne's 'O my America: my new-found-land,'[19] as an image for the delighted sense of discovery in exploring a desired body, is more suggestive of awe and mystery when we remember that America and Newfoundland had been discovered less than a century ago and were not yet settled; indeed, the image is one of 'virgin territory'.

It is useful to know a little about ships and the sea in the past – if only to have looked at a few pictures – to understand better many narratives and images; perhaps especially to have some awareness of hardships and dangers now eliminated. Wyatt's *The Galley* exemplifies this source of imagery.

We should pick up as much knowledge as we can of places, not only their locations, but their associations:[20]

> And all who since, Baptiz'd or Infidel
> Jousted in *Aspramont* or *Montalban*,
> *Damasco*, or *Marocco*, or *Trebisond*,
> Or whom *Biserta* sent from *Afric* shore
> When *Charlemain* with all his Peerage fell
> By *Fontarabbia*.

These were real places; it is well to know that, for instance, Cockaigne, Avalon and Lyonesse are not. Difficulties sometimes arise because place-names have changed; even whole countries have, as named entities, changed names or shapes, appeared or disappeared. Byzantium becomes Constantinople, then Istanbul. Even an hour spent with, say,

Muir's Historical Atlas[21] is better than nothing, at least showing us how much we do not know.

8 The student of drama must know something of the history of the theatre; not merely about famous actors and actresses, but – more important to literary studies – the great changes in the shape of theatres and stages, the use of scenery and lighting, the social status of drama, theatrical practice; we should know, for instance, that all Shakespeare's female roles were originally played by boys and that he had to create his scenery almost entirely by evocative words.

9 The more general information we have, the better: the more we know of events, such as Marathon, Waterloo, the fall of Constantinople, the Jacobite risings, American Independence, the Great Exhibition; of famous people such as Pythagoras, Newton, Catherine the Great, John Wesley, Florence Nightingale; and of allusions, whether to the Wise Men of Gotham, the *ignis fatuus*, Limbo, or Will's Coffee-house. A good vocabulary concerning past times – garments, weapons, heraldry, customs, occupations and everyday objects – is useful; even a quick look at pictures, as in an illustrated history of costume or houses, is some help. We shall never carry in our heads all the details of times past, but it is as well not to picture a culverin as a sort of spear or a Geneva gown as a pretty Swiss frock.

10 A first-degree student need not worry about such details of bibliographical description as a student once told me looked like knitting-patterns, but should learn a little about the history of books, printing and the profession of authorship, so that he does not imagine Jonson's plays were reviewed in next day's paper or that a Victorian 'three-volume novel' was necessarily very long.

11 The more a student knows of the history of ideas, the less likely he is to read into literature implications that could not possibly be there. Locke's *Essay Concerning Human Understanding* (1690) is illuminating in relation to *Tristram Shandy*

(1760–7); but George Herbert's 'though the whole world turn to coal'[22] cannot be an evocation of the Carboniferous Period, as a student once told me, since that concept, and even the time-scale in which it is contained, were unknown in Herbert's day.

Clearly no student, and no professor at retirement age, is going to know everything it is desirable to know; but the better our knowledge, the fuller our enjoyment. Finally, though often a pleasant and profitable fast reading for a general impression does not demand exact understanding of every word, in one situation the student must pursue the meaning of every single word, understand every passing allusion: that is when he is studying a set book, especially any set book on which there will be context questions.

13

INSTRUCTION AND DISCUSSION

> I must go beat my brains against a bed-post,
> And get before my tutor.
> Thomas Middleton, *A Chaste Maid in Cheapside*[1]

We pursue literary studies mostly in silence and solitude. The essential activity is reading: close, attentive, thoughtful, copious reading. If a student is able, keen and persistent enough, and has access to the necessary books, he can reach degree standard virtually in solitude; thousands of people have, by difficult self-discipline and perseverance, achieved external degrees by correspondence courses, sometimes carrying heavy personal handicaps, or supporting themselves by paid work at the same time.

So is the enormous expense of university staffing, lecture blocks, a tutorial system, audio-visual equipment, necessary? For scientific studies, laboratories are needed; but for literary studies, might it not be better to provide just very good libraries, warm and well lit, open from seven in the morning until midnight, with plenty of working places, extra copies of the most essential books, library staff qualified to give students occasional advice as well as to look after the books, and a rule that any student breaking the silence or doing damage, except by mishap, should be excluded, fined and put in the stocks?

If we ask, 'What are academic staff for?' the full answer is not, 'Instructing students'. The function of a university is not only to *transmit* knowledge; if it were, the science laboratories might today be dedicated to alchemy. The other great function is to *extend* knowledge. Part of the business of a university is to support a community of able, highly trained, keen specialists who can collect new information, produce fresh ideas, and test them by putting their heads together. (If they sometimes do the latter with noisy clashing of horns, this too may serve scholarship; intellectual progress is made partly through sharp debate among equals.)

However, instructing students is part of the work of academic staff; and there are several functions of organized instruction.

Frankly, one function is to reinforce our own often patchy self-discipline. The student sincerely intends to study Book I of *The Faerie Queene* this week; but there is TV in the common-room; friends want to go to the pub; he is in love; there is a crisis in the Students' Union, a match, a party, a production, a fit of the blues. The student means to do the reading, but it gets put off while . . . until . . . because. . . . If it has to be done in order that he may not look a fool in the class at 10 a.m. on Tuesday, the student's immediate motive may not be the best, but the reading gets done. Having to produce some kind of written exercise forces the student to organize vague impressions into something sequential and articulate. A low mark or an acid comment may warn him that the standard he is attaining at present will not do. Not all bad work is sheer laziness or stupidity; sometimes we honestly do not realize what better work could be done, until someone else shows us. And it is one thing to drawl out a dogmatic opinion of, say, the fatuity of Collins's *Ode to Evening*, over a pint; quite another to be forced to justify or renounce that opinion by a tutor with polite Socratic questions and an ironic eye, or even by a fellow-student in class with a vehement contrary view.

However, even the student who is so devoted that he has to

be coaxed away from work for some fresh air, and such a striver after exactness and honesty that his work can scarcely be faulted, has much to gain from tutors and from contact with other students.

One function of instruction is inspiration, excitement, stimulation. A lecturer who is on fire with the intellectual excitement and emotional power of the subject can scatter sparks. I have sat in lectures where the thrill flashing through the room could be felt like an electric current; I have had tears running down my face; I have heard an undergraduate say after a superb lecture, 'We *did* ballads at school, and I thought they were dead boring . . . but this – this is a knock-out!' A friend of mine was allowed into a lecture as a visitor, and all the way back to my house kept saying, 'That was an intellectual feast . . . a real intellectual feast!' There were moments of inspiration or intellectual revelation in lectures I heard in 1941–4 that are still part of my mental treasure some forty years later.

Not all lectures are like that; and probably if they were we could not stand it. Another function of instruction is to save time. On some subjects, such as 'Medieval Background', or 'Developments in Prose Style in the Seventeenth Century', or 'The Experimental Novel', a competent instructor, in a course of lectures or classes, may give students a mass of useful material *selected*, *organized* and *clarified* so that it is much better equipment than a student could gain for himself by desultory, unguided, probably ill-proportioned and often perplexing reading.

Instruction provides counsel and correction; the student has opportunities to ask about something he cannot quite sort out for himself, when he knows what he does not know; and a tutor can also show him that he did not know something properly, when he thought he knew it. Though it can be embarrassing to have our mistakes pointed out, it is obviously sometimes needful if we are to make progress.

But probably the most important of all functions of organized instruction is that it brings each student into contact with other minds. We need much solitude if we are to do worthwhile study; but we also need to remember that in the intellectual life, as in the laws of mechanics, friction is a necessary part of motion. By listening to various opinions, by rubbing against other (and often better) minds, we test our own views, perhaps modify them, see a subject from a fresh angle, learn new questions to ask and, perhaps, new answers. This is the great educational advantage of a resident course over an external degree. Incidentally, those engaged in research sometimes share ideas or findings not yet available in print.

There are three main types of university teaching, though names vary: the formal *lecture*, in which someone of exceptional specialist knowledge of a branch of the subject discourses uninterrupted for about fifty minutes; the *seminar* (class, discussion group, colloquium, conference, symposium), in which it is hoped that a group of students will freely discuss a topic – often after one student has read a short paper – under the guidance of a specialist; and the *tutorial*, in which one or two students spend an hour with a specialist, enjoying a closer, more specific personal attention, or possibly not enjoying it.

The lecture has been somewhat out of favour lately, wherever 'participation' has been a shibboleth; some think 'discussions' preferable as being more 'democratic'. Certainly the free exchange of opinions is not only a vital civil liberty, but vital to full intellectual development; and taking part in a discussion forces us to construct and articulate ideas; on the other hand, ten of us articulately pooling our stock of ignorance are hardly going to achieve much knowledge. One good lecturer can give two hundred students simultaneously a starter kit of knowledge; on the basis of that plus the reading advised, the students have something to discuss, at a level on which their different approaches contribute something meaningful.

Difficulties may be caused by lecturers who lecture rather badly, or by students who do not know how to listen.

School teachers have to have lengthy training for teaching; university teachers have little or none. Some might benefit from a short course on technique, notably voice production; but school teachers need special training because many of their pupils will not want to learn, or will have little intelligence; so teachers must learn to control, stimulate, simplify, cajole, jolly along, encourage and occasionally intimidate; a student who does not positively want to learn has no business in a university. If a university teacher has to think at all about a teacher's problem of 'discipline', 'class control', there is a seriously immature student around.

It is fair enough to say that someone paid partly for lecturing ought to do it properly. Unfortunately, outstanding scholarship does not automatically endow anyone with lecturing skill; moreover, the devoted scholar is often a shy person. Students might be surprised sometimes if they knew what an ordeal a lecture could be for the lecturer; one of the most compelling, profound *and* entertaining lecturers I ever heard suffered agony before every lecture, even after thirty years' experience, and was chalk-white whenever she walked in. This is not a book for lecturers, but there are some things a student can do to improve a course of lectures:

1 We usually hesitate to occupy the front row. Lecturers do not spit venom. If a lecturer's vocal power is poor, it is common sense and kindness to sit as near as possible. The lecturer will feel less strained and probably speak better, with a compact group and no needless distance.

2 It is brutal to yell 'Speak up!' to someone who may be already fighting twanging nerves and a dry larynx. It is reasonable to put a hand up early in the first lecture and politely mention that some students cannot hear; not all

inaudible speakers are aware of the difficulty. Possibly a microphone may be available. To humiliate or upset a shy lecturer will not help; we all know what happens if we keep saying to a clumsy child, 'You'll drop it!'

3 Give feedback. Fidgeting and open displays of boredom make bad worse. Do the lecturer the courtesy of trying to follow; let any positive response show, take what notes you can; ask questions where appropriate; those who honestly try to listen do learn at least part of what the speaker is trying to teach.

4 Do not assume that all the fault lies with the lecturer. Real attentive listening is quite an art. (Nowadays some of us damage our capacity for real attention by such habits as having a radio on when *not* listening to it.) A friend once told me how a student said, apparently in a tone of pleased astonishment, 'I used to think your lectures too hard for me, Miss H——, but this term I've read the books you told us to read, and I can follow everything you say!' We must not expect everything to be baited or sugared for us as if we were still at school.

5 The art of taking notes in lectures has to be learned by experience. If we try to take everything down, we miss the fourth sentence while trying to get the third down. Notes should be a fast précis of the lecture, plus exact details of any title, date, reference or fact that we may need for the follow-up reading. It is wise to leave some spaces in notes for additions and corrections. Frantic indiscriminate note-taking may produce something almost unintelligible a week later; good notes are a lastingly useful reminder of what was heard. The effort to make useful notes is as much a help to concentration and following the sequence of ideas as compulsive ballpoint-automatism is a hindrance.

The essential to make a seminar (or tutorial) a success is to do the preliminary work, whether this is a quiet reading of two books for comparison, or taking one's turn to write a paper to read. Without the basic knowledge, we not only cannot say anything useful in a seminar, but cannot assess, or even understand, what other people say.

I cannot honestly say that a second essential is that tutor and students shall all contribute something, each taking a fair share: a seminar can be memorable and valuable because the tutor, or a keen and informed student, was fired to talk eagerly for twenty minutes, making an oustandingly able contribution; but in general a wide and 'fair' sharing of the discussion works best.

It is important that the tone of the discussion should be thoughtful and civilized; a seminar is not a battle or football match to win or lose; everybody wins if everybody learns something; over-eagerness to score points, as opposed to readiness to contribute ideas and evidence, may tempt us to intellectual dishonesties.

A seminar is also not a factory shop-floor. In the trade-union world, occasionally it really can be selfish to work so hard as to take work from others or set standards killingly high; but the students who erected as a principle for seminars, 'Whoever speaks more than twice is a creep' had not evolved sociologically beyond the fourth form. Even competitiveness is healthier than a concept that rations enthusiasm and resists interest.

Though friction is necessary to movement, though minds make sparks when rubbed together, friction is not necessarily abrasiveness; machines function better when oiled. (No, this is not a recommendation to take a drink before a seminar!)

Two great concepts of university life that are no less valuable for having become clichés are: '*the courtesies of debate*' and '*the fraternity of scholars*'. They cannot be practised to perfection; we are all imperfect; but they represent essential ideals of good academic life.

Some American students reacted to a lecture with which they disagreed by chanting 'Wimsatt! Wimsatt!' as much as to say, 'You haven't studied Wimsatt's theories, which render what you are saying obsolete.' This was at least relatively intelligent rudeness; Wimsatt is a critic and was relevant; but any discipleship so exclusive, any adherence to a One Theory that Explains Everything, has an unpleasantly totalitarian flavour. W. K. Wimsatt's *The Verbal Icon*[2] is worth reading and includes some useful warnings against imperfect critical approaches; but if the students who were turning an honourable scholar into their academic *Führer* knew all about Wimsatt already, they might expect to learn most from a lecturer with a different approach.

There is an intellectual ministry that is generous, sensitive, creative: over Britain and much of the world, academics are daily trying to help students to know, to understand, to think clearly for themselves. Such a ministry demands both a real respect for the subject being studied, and another kind of respect for the students; an experienced guide leads novices where Alps on Alps arise, and helps them to learn mountaineering. Intellectual priestcraft is fortunately not very common in British universities: an excessive desire to nurture not grateful yet independent graduates, but uncritical, narrow disciples. The dogmatic panjandrum may still have valuable ideas, and his students should pay attention to them; but if that régime is somewhat dictatorial, they should take special care to travel in some other countries of the mind.

I heard of a student who went to a seminar on Marxist literary criticism from a commendable desire to learn something about it; but some convinced Marxist students there dogmatized so hectoringly, so scorned him as a naïve middle-class dodo, came so near to threatening him, that after two sessions he stopped going, having gathered only that all Marxists were rude and blinkered bullies – which is far from true. Proper answers to his queries, patience and good humour, might even have converted him. For a student of nineteen to be sure he is right is even more absurd than for the scholar of fifty-nine, who has at least done a lot of work on the subject.

We should see through the popular fallacy that a harsh, aggressive manner is evidence of sincerity, some sort of honest no-nonsense plainness. It is far more likely to be an aspect of egotism, disregard for the feelings and dignity of others. Academic debate should be oiled by plenty of unselfish tactful touches such as, 'Yes, but . . .', 'Are you sure that . . .?', 'That's true, but . . .', 'Do you agree that . . .?', 'Isn't another possibility that . . .?', 'But are we certain that . . .?', 'But what do you think of . . .?'. All such politeness and restraint is only a way of acknowledging that other people have minds too, and that this all-precious creature I alone call 'I' is capable of sometimes being mistaken!

Courtesy need not always be verbal; there are looks and tones and smiles that keep discussion sweet. It was once my privilege to attend a series of seminars where, rather unusually, a distinguished scholar sat at each end of the table. The two often disagreed; and then how they went for each other! They also had a notably generous and caring friendship; and while the clashing rapiers showered us with illuminating sparks, the glow of true respect and affection in the eyes of those scholars filled the room with warmth.

Scholars, though disagreeing on points, should be a fraternity, and students should be younger brothers (or sisters) in that fraternity. As we impose our various egotisms on others – all of us sometimes; some politicians, agitators, advertisers and bigots most of the time – avoid or suppress truth, perhaps spread lies. Scholarship is concerned with the pursuit and spread of truth, aiming, at least, at the disinterested pursuit of truth.

Those who have struggled as far as even some care for accuracy and intellectual honesty and some intelligent love for the more demanding arts are in a minority even in a civilization as expensive, rich in opportunities, and free, as our own. Posters soon show evidence that some people positively like to deface what others design; probably most schools have their hobbledehoy Hitlers and Hitleresses who evidently enjoy making gifted pupils miserable; millions not spiteful, usually lovable, in many ways good, aspire chiefly to possessions and bodily pleasures, or to a culture without intellectual rigour, consisting of pop music, light romances, thrillers and televised situation comedy. The trendy use of judgments and facile modishness), but a pressure upon people warning against intellectual snobbery (which is bad not only for the soul, but for the intellect itself, leading to narrow judgments and facile modishness) but a pressure upon people to feel guilty about doing anything properly.

In many countries students and scholars are in a threatened minority; even in Britain physical attacks on students, simply for being students, are not unknown. In some sense all sincere scholars and students are keeping the lights on, and trying to intensify those lights, against the mere natural darkness of ignorance and the deadly smogs of obscurantism, intolerance and barbarism. Handicapped by all our own embarrassing inadequacies, we are in this sense a noble company engaged in noble work. Sometimes we might try a little harder to show one another the courtesy appropriate to such a company.

14

A FEW FAVOURITE FALLACIES

Knowledge is proud that he has learned so much;
Wisdom is humble that he knows no more.
William Cowper, *The Task*[1]

Every human being is from birth programmed for fallibility in all fields, not excluding even chosen specialist disciplines; our first step towards wisdom is to recognize this. We can, however, make the best of our imperfect equipment and avoid further programming ourselves for gross and avoidable blunders. If I keep returning to the idea that humility is one of the keys to scholarship, it is because most of us (including me) need frequent reminders of this; that, as Thoreau startlingly says, 'The works of the great poets have never yet been read by mankind, for only great poets can read them.'[2]

The first and worst mistake we can make about a book is to hold an opinion on it before we have read it. I have seen someone hand back a copy of a book with totally dismissive contempt – and the pages uncut. In a clever and enjoyable detective story, the hero, ashamed after losing his nerve, apologizes: 'I acted like Little Lord Fauntleroy',[3] because many people think of Lord Fauntleroy as a spoilt little cissy in velvet; but all the stereotype has in common with Frances Hodgson Burnett's novel is the hero's velvet suit; her Little Lord Fauntleroy is a brave, unselfish, self-controlled little fellow; and though the story, intended for children, is didactic and sentimental for the fashions of this decade, its values are largely permanent and it is not without insights or even intelligent ironies.

We have all also a disconcerting ability to read-and-not-read. I have not only read all Byron's poems, but studied the best of them, I would have said attentively, and for several years gave an annual course of lectures on Byron; but, asked one day in a quiz which American pioneer was mentioned in *Don Juan*, I found I had no idea – not the faintest stirring of memory, not even an educated guess. The pioneer is Daniel Boone, shown as 'General Boon, back-woodsman of Kentucky', as an innocent hunter in the wilds, contrasted with corrupt civilization; and Byron gives him seven stanzas, not just a fleeting mention![4] Moreover, the contrast of a wild, harmless child of Nature with the bloody depravities of politics and 'civilization' is a characteristic of Romantic literature such as lecturers point out. Did I repeatedly turn that page when half asleep? I do not know; but no student can be confident that he will not do much the same one day; and this is a warning not only against inattentiveness, but against sweeping judgments on what we think we have read. *Have* we read it?

We should be careful about critical terminology. We seldom misuse minor technical terms of narrow definition, such as *amphibrach*, *cento*, *exemplum*, *malapropism*, *poulter's measure*, *zeugma*; we are not likely to use them unless we know what they mean. Some terms are harder to define and do not always mean the same to all critics, for example: *classicism*, *decadent*, *humanistic*, *image*, *myth*, *realism*, *romance*, *romanticism*, *sentimentality*, *symbol*. We should, therefore, be sure that we are understanding such a term in the same sense as the critic; or that, using it ourselves, we place it in such a context that the sense in which we are using it is clear. For instance, in some translated Marxist criticism we may find *humanistic* used in a sense so far from that in which it may be applied to Roger Ascham that we have two different concepts. *Romanticism* is impossible to define in a few words, and is not something self-consistent. *Decadence* may refer with reasonable precision to a minority literary movement of the late nineteenth century led by such as Baudelaire, Huysmans, Symons, George Moore, Dowson, Wilde; refer more generally to the last stage of some literary fashion or period; or be mere peevish abuse in some jeremiad.

We should be especially careful when considering a literary work as a representation of life, talking about its 'truth'. First, it is not usually the task of a literary work to represent 'truth' in a documentary way; apart from such things as obvious allegory, myth, fairytale, vision, caricature or satirical extrapolation, there is room for less overt fantasy, idealization and use of symbols. Any fictional portrayal of life is bound to be fragmentary and selective; and any selection has some kind of bias. We must not forget that 'imaginative literature' is precisely that – a construction of the imagination; or that aspects of fantasy may meet real needs of the psyche deeper than the need for amusement and consolation.

For instance, from one point of view much love poetry is the expression of besotted idealizations; but love poems may also be seen as verbal presents made in tenderness, a refinement of wooing, an offering of skills to one's beloved; and on another level of experience some apparently extravagant, almost idolatrous love poetry may be a kind of myth-making, an attempt at conveying strange inner experiences that are genuine; Dante's imaginative treatment of Beatrice is the most impressive example.

Second, we must constantly try not to measure all 'truth' by the little we have seen of life – little, in proportion to the multiplicity of life, even for the centenarian. (So far, in growing older, I have found myself less and less eager to stigmatize something in fiction as 'impossible' or 'totally untrue to life' as I see more and more of the astounding range of human possibilities.)

Third, we should remember that 'realism' is a rather ambiguous term. Writers classed as *realistic* are often those who deal extensively with the unpleasant aspects of man and society – Maupassant and Dreiser are examples; but it is possible to treat of any aspect of experience with the attempt at detailed, objective, truthful presentation that constitutes realism. Lice and excrement are no more and no less *real* than soap and marmalade; the humiliations and bitterness of a wrecked love affair are neither more nor less real than the gratitude and irradiations of a happy one.

Similarly, we should be cautious in our use of the term *sentimental*. It has a genuine critical function, as in the fairly exact term *sentimental comedy* (e.g. Steele's *The Tender Husband*); but we should not use it in easy contempt for emotional attitudes we do not happen to have experienced. As good a definition of sentimentality as any is probably Wordsworth's 'Affecting more emotion than I felt . . .'.[5]

Because few people can now contemplate the infant Jesus with the marvelling certainty of Crashaw, we are not to find insincerity in his *Hymn of the Nativity*, with its wealth of lovely fancies that to Crashaw contained theological as well as emotional truth. Plenty of people share my subjective view that childhood is far more unhappy than adulthood; but this does not mean that Vaughan is hypocritical or self-deluding in *The Retreate*. Patmore's *The Angel in the House* may now in places seem outdated to the point of quaintness; that need not invalidate it as a study of genuine emotions in a social framework different from our own.

I think male students may, marginally, be quicker to suspect 'sentimentality' than most female students; and possibly a majority of women students may be more ready to reject a work because it is 'cynical', because it expresses some of the negative feelings virtually all of us sometimes have, or treats of some aspect of the unfairness, pain, mess or perplexity of life; yet these exist and to articulate them is a legitimate function of literature. Hardy's 'Who holds that if way to the Better there be, it exacts a full look at the Worst'[6] does not define a man without hope or benevolence. *The Hollow Men* is a recognition of real experiences of emotional failure, spiritual aridity; but even in lamenting the condition it implies other possibilities; Rochester's *A Satyr Against Mankind* is not recommended to cheer us up, but forces us to ask ourselves some of the questions we need to face; and sometimes the sad stoicism of Rochester, Housman or Hardy may keep us better company through trouble than something more cheerful, but shallower. To write with such controlled art of man's indignities is at the same time evidence of human dignity.

Cynicism is in a sense the obverse of sentimentality; and a too facile dismissing of something as either cynical or sentimental may be a symptom of our own limitations: a tendency to evade the full pain of reality, or the subtler self-protections of the emotional cripple; one may be unable to stomach Zola's *La Terre* or even Margaret L. Woods's *A Village Tragedy* for the bestial but all too credible degradation shown in the great novel, the quieter respectable cruelty in the fine minor novel; but many another reader may shrink from, say, *Pearl* or *The Franklin's Tale* or *Middlemarch*, from Esther Summerson or even Desdemona, much as Satan stood abashed before Ithuriel and Zephon:[7]

> And felt how awful goodness is, and saw
> Vertue in her shape how lovly . . .

Generations of students have had to be warned of the prevalence of ambiguities and ironies in literature. These do provide some of the richness and excitement of literature. However, once we have been well trained in this respect, we should also be careful not to hunt out subtleties that are not there. Milton's Satan, feeling 'how awful goodness is' felt it only as 'awe-inspiring'; that he found it 'insufferable', like

Oscar Wilde contemplating Little Nell, that Milton was hinting a schoolboyish suspicion of priggishness, is quite out of the question! Thomas Campbell's:[8]

> O Love! in such a wilderness as this,
> Where transport and security entwine

is certainly not using *transport* ambiguously; it here has its sense of 'rapture' and no other – nothing to do with safe travel. We should not, in digging hard for buried meanings, lose sight of the surface landscape: *Moby Dick* is a story of character and of adventures hunting whales; *The Ancient Mariner* is a ballad of supernatural marvels; *Huckleberry Finn* is a tale of a young scapegrace's amusing and exciting adventures; they contain a very great deal more, but we should always start with the obvious.

Sexuality does pervade life and often affects our motives in ways we do not ourselves recognize; at different times various aspects of it have been tabooed and driven underground; so we do well to keep an eye open for veiled sexual allusions; but we should not become so obsessed as to see them where they are not. Blake's '*I saw a chapel all of gold . . .*' with the serpent rising between the white pillars of the door, almost certainly has a sexual meaning, though precisely what is disputed; it could well, for instance, be about callous conventions and crude ideas of marital 'rights' defiling innocent sexual joy. But if we then try somehow to decode Milton's Samson between 'the massie Pillars With horrible convulsion'[9] as an image of sexual guilt and its destructiveness, we shall merely make fools of ourselves.

We are all to some extent the creatures of our social and economic situations; but, again, to hunt for class-consciousness and class-struggle, women's-lib themes, or other comments on the social order, in everything from *Beowulf* to *The Whitsun Weddings* will lead us into weirdly perverse misreadings. (A commonsense corrective may be to ask ourselves how much time the average person spends on thinking about politics and society, as opposed to immediately personal worries about pay, prices, or employment.)

If I go to bed hungry and dream that I am eating sausages, it is unlikely that the dream springs either from sexual frustration I dare not admit to myself, or class resentment that I can afford only sausages, not steak; it is far more probable that my 'imagination bodies forth the forms' of sausages because I want to give them a 'local habitation' in my aching stomach.

15

SOME RUDIMENTS OF STUDY TECHNIQUE

With much ado, his Book before him laid,
And Parchment with the smoother side display'd;
He takes the Papers; lays 'em down agen;
And, with unwilling Fingers, tries the Pen:
Some peevish quarrel straight he strives to pick,
His Quill writes double, or his Ink's too thick. . . .
No more accuse thy Pen: but charge the Crime
On Native Sloth, and negligence of Time.
Think'st thou thy Master, or thy Friends, to cheat?
Fool, 'tis thy self, and that's a worse deceit.

John Dryden, translation of the *Third Satyr of* Aulus Persius Flaccus[1]

First, the new student can add one more to Charles Lamb's collection of 'Popular Fallacies',[2] soon refuted: the myth that the Good Scholar is Good for Nothing Else. Discontented students affect to despise better adjusted students: Swot, Gnome, Grind, Bookworm, Creep, Grot, or worse. There is a kind of folklore image of the conscientious student: a peering, hunched, unsmiling, unsociable, bloodless, sexless, cocooned, dusty biped, narrowly and selfishly preoccupied with his books alone. A valuable 1978 student guide included an essay – all too brightly written, neatly argued and not devoid of sensible points – advising students that a First is not worth getting; better aim at a Second with some worthwhile extra-curricular activities.[3]

Now, if there were a chalk-or-cheese choice between graduating with first-class Honours and graduating as a vital, versatile, affectionate human being, I would unhesitatingly advise any student to opt for the latter.

The only snag is that the dilemma is imaginary. It is like the most amiable of radical slogans, 'Make love, not war', which, though it contains a modicum of moral and psychological truth, does not present mutually exclusive alternatives.

From a philistine backwater and over-restricted life I stumbled into university conditions, in wartime, with inadequate food, cold, six hours' compulsory war work weekly plus firedrills and so forth; and it seemed like entering a heaven of interests, friends and choices. I probably spent more time on political activities (half but not all of which I now regret) than on study; the joys of friendship, the pangs of love, religious perplexities, seeing good plays and films, my own writing, and hours and hours and hours of sheer educative *talk*, mostly not about literature, all claimed my time and energies. I still won my First. I recently photographed in her new doctorate robes a friend who is also half of a very good marriage and widely popular; she loves music, can cook admirably, grow her own vegetables, decorate a house, re-cover chairs, tile a bathroom, help lame dogs, manage Siamese cats, make curtains, jam or witty conversation. She is about to edit her third scholarly text.

Naturally I do not go about asking everyone I meet what class they got; it is neither my business nor very important; but as far as I know I have never yet met a first-class graduate, or any university teacher, who did *not* pursue some other interests, know something of a wide range of subjects, do some services to others beyond the obvious line of duty and maintain some real affections. The exhausting bores and disquietingly shrivelled hearts I have so far encountered have almost always been among the ignorant.

One last reassuring example: I knew a young man who

obtained a First (mainly in Philosophy) after three years in which he was notably active and respected in left-wing politics and in the NUS; had many friends, and a full social life; and cultivated his other great interest, music, with several minor interests. He is now a university teacher with a number of publications to his credit. All this he achieved with no eyes.

The secret of successful study is not to bury oneself alive behind a wall of books, excluding everything else; this would be absurd for, of all things, English, which depends so much on broad general knowledge and interest in emotions and motives; but to spend enough time daily in *effective* study. One hour spent really reading *Prometheus Unbound* is worth more than three hours spent pretending to read it, half asleep in front of the fire with the radio on.

Psychologists have given some attention to study technique, and a good book on the subject is likely to give any student a great deal of help. Several such books will be listed later. A few paragraphs, however, suffice to mention some simple commonsense principles that by no means all students have grasped before starting their courses.

We should consciously plan our use of time. Some students are greatly helped by making formal, written time-tables, filling in lectures and classes first, then periods for private study and for reasonable extra activities, not forgetting to allow time for trivial necessary tasks – cleaning boots, washing hair, shopping – and seeing friends. We do not want to plan our lives in such a way that all human contacts become 'interruptions'! Certainly the student who says, 'I never seem to get down to it!' may find that a week's analysis of where his time is going, followed by a strict time-table for the next week, is excellent therapy. Other students find close time-tabling needless or even counter-productive; but every student should at least consider frequently: what are the tasks to be done? how long do I need for them? when shall I do what?

Self-discipline is never very easy; every course of successful study, like every life even half-way decent, is testimony to hundreds of little self-conquests. We often have to tell ourselves firmly, 'I will not do so-and-so – until I have done this-here.' (I still keep a few books sealed into bags, with the stern label, 'Reward only' – that is, not to be looked at until I have written an article, or dug the garden, or whatever I am currently finding it hard to complete.)

The student who has merely set himself firmly down where he intends to work and seized his book, notebook and anything else he needs, has won the first victory. The next struggle is for continuing concentration. This comes naturally to few; we find it hard to achieve the intense, exclusive attention of a cat at a mousehole, precisely because Man is such a versatile animal, capable of being interested in many things in one hour.

And concentration on the work is not continuing to look towards the book; it is continuing to read it with full attention.

There is no sense in making our task harder by avoidable causes of distraction. We should do what we can to keep our surroundings quiet. One of the mistakes of our epoch is the widespread addiction to a kind of non-experienced entertainment: records, radio or even TV in the background of conversation or activity. 'Music while you work' may have its value when people have to do monotonous and almost mindless work; for serious study, any such background takes off some percentage of that attention it is so difficult to fix anyway.

Usually a student can concentrate best on study when sitting on a fairly hard chair, at a desk or table, not slouching too much; and in a room that, while not too cold for comfort, is not so warm as to produce lassitude or even sleep; with some ventilation (brain functioning depends on oxygen!) and enough light.

Students have overcome ghastly illnesses or grave handicaps; but some of us make handicaps for ourselves; we should attend to any health problem as effectively as possible; wear spectacles if we need them; eat a diet as healthful as grants and our other conditions permit, especially with enough protein and enough roughage – constipation does not stimulate the mind; avoid over-tight clothing or uncomfortable shoes; avoid dependence on alcohol, tobacco or any unprescribed drug; get some exercise daily and enough sleep nightly. This sounds grannyish and boring; but it is more efficient and more fun to be fit.

It is well to keep distracting objects or daydream triggers – worrying letter, newspaper, magazine, sweetheart's photograph – off the desk; and it saves time if objects likely to be wanted – paper, pen or ballpoint, ruler, dictionary – are on the desk before work begins.

After that, it is a matter of painfully fetching one's attention back, like an untrained dog, every time we catch it wandering. One cheering fact is that our span of concentration does improve with practice.

We do need now and then to take a break, though we must try not to confuse *need* with *want*. A break after an hour's genuine concentrated work is reasonable; but five or ten minutes is enough. The best break is often a little light exercise; we can get stiff sitting. To do a small puzzle refreshes my brain; a few minutes of music suits someone else; often, if we feel tired and the mind is jamming, we can unjam it by doing some other small task we have been putting off: sew on that button, write that note of thanks, wash those socks. To have a break that is also a tiny victory is good for morale.

And we have to check, and keep on checking, that we *understand* what we think we are *reading*. This is one reason why it is worth while to make notes; and why notes should *copy* only quotations we think it would be useful to learn by heart, but give the *gist* of everything else. We must bear in mind all the time, too, that, except for the immediate self-discipline, notes are no use if next month we cannot understand or even decipher them. Revision for examinations will make heavy demands on our notes; there will have been many books we could read once only; if we remember that in making notes we are stockpiling ammunition for our last weird battle in the west, we shall make them with care.

Students often need to learn things by heart: useful quotations, dates, facts, arguments and so on. It is much easier to memorize something we understand, so we should aim at a thorough understanding first. It is much easier to learn something as a whole – a sonnet, a paragraph, a psalm – than to memorize it a bit at a time, when the bits are apt to get out of order, or one bit falls out while we try to pick up the next. And we have to remember the educational principle already understood by Chaucer: 'me semeth betre to wryten me un-to a child twyes a good sentence, than he forget it ones.'[4] A very few people – Macaulay is said to have been one – can remember anything they have read once; most of us can memorize only by re-reading and re-re-reading and many-more-readings.

It is very easy to think we know something by heart when we are only tired of repeating it. The test is to put it aside, do something else, and try again in an hour. We may find we have already forgotten the piece. If so – back again to perhaps tedious repetition, until the thing seems really fixed in the mind; and test the learning again after a day or two. Many actors learn their parts last thing at night and check again in the morning; this may help.

An important reason for keeping up with our work and not leaving much to do near to the examination is that, in general, memory works on the rule of 'last in, first out'. This appears pathetically in very old people, who can remember some childhood treat or task but not where they put their spectacles ten minutes ago. What we have just 'learned' is not yet 'fixed', and may slip away when we need it. Besides, we need material not for an examination only, but probably for discussion in classes.

One other point should be made about the study course in general: university terms are short, and vacations are not for vacuity. Especially in a subject such as English, with a huge programme of actual reading, it is assumed that a substantial part of each vacation will be spent on reading and consolidating knowledge. Probably the best service parents can do for a student is, if possible, to provide an undisturbed corner for vacation study. (Even today, some parents need reminding that a woman has to work through the same syllabus and get through the same examinations as a man; it may be necessary for Janet to give some help in the house, but for student Janet to do all the helping while student John can get on with his work is outrageously unfair.) Less fortunate students may be able to find some quiet in a library, a friend's or relative's house, staying at college, and so on. It is a pity that many students have to seek paid vacation work; this should be kept to a minimum; but the student of literature can at least keep relating what he sees of some new community to some aspects of his reading; literature is about life. (He had often better do this inwardly and secretly, or his attitude may cause resentment.) Similarly, occasional holidays are highly desirable, but that book you have kept meaning to read can be packed for a rainy day, and almost any holiday will yield some observations relevant to something in literature. We can keep an eye open for useful matter, without being obsessed bores.

Moreover, the more work a student can manage to do in the vacations, when very possibly social and cultural opportunities are not so varied, the more time he can spare during university terms for the excitements university life can offer.

16

ESSAYS AND PAPERS

> I like a good grip; I like to feel something in this slippery world that can hold. . . .
>
> Herman Melville, *Moby Dick*[1]

Prescribed writing tasks, whether a short weekly essay presented to a tutor, a more ambitious paper presented to a seminar, even a small dissertation as part of course requirements, demand different levels of detail; but all fulfil three obvious educational functions.

A weekly or less frequent task has a benevolent disciplinary function. Almost all of us can easily fritter time away for days on end, especially in an environment full of interesting alternatives to work; the task ensures that each student does at least some organized work; it also enables someone else to judge the quality of the work, and so to help the student's progress by giving appropriate encouragement, correction and advice.

Having to write an essay forces a student to read more attentively and questioningly. The situation may seem comically artificial: one of the few things certain about Shakespeare's motives is that he did *not* write *The Comedy of Errors* envisaging a student trying to write something perceptive about the comedy while a tutor had a red ballpoint ready for the errors. Yet real attention is such a difficult art that most of us need coercive situations to train us.

If possible the student should think about an essay on a

second, or *n*th, rather than a first, reading of a primary work; the first reading should be relaxed; pleasurable on the level of pleasure requiring relatively little effort; and aiming at a general, interested personal experience of the book; in fact, reading an imaginative work much as the author would expect. To start reading *Lord Jim* consciously looking out for material for an essay on 'Conrad's treatment of the concept of honour' and noting relevant bits to quote, may destroy the impact of the novel as a whole, as an exciting, moving fiction, and in hunting for bits that serve our turn we may miss almost everything else worth noticing. So, ideally, each first reading of a book should aim at a general survey, a unified experience, and notes for an essay be made on a later reading.

The primary work should be treated as far more important than any critical study. Unless the essay topic requires discussion of a specific critical view, a competent student should be able to make a passable essay out of no material but the imaginative work itself, read with real attention, plus any information needed for ordinary understanding. However, reading good criticism does improve our literary judgment, and the student should try to read all the studies recommended. Though he may not draw much on them for the essay, they are likely to be quoted in discussion later. Ideally, the essay-haunted student reading a critical work will, again, read once for a general grasp, again with a view to the specific essay; time will not always suffice; what is a mistake is to rush through a critical work with ballpoint ready, grabbing at quotable sentences without following an argument.

The third function of the set writing task is to force each student to articulate some coherent opinions; not just to have a vague experience of a book, or, as we all find so easy, to think that he has thought about it, but to test his thinking by constructing an exposition or arguing a case. We can be very confident that we understand something – until we try to explain it to someone else.

Even the most generously flexible essay topic, such as, 'For next week, write something about Traherne', is not an invitation to the student to put down a few facts and some unexamined impressions about Traherne, add a paragraph from a critical study (with no inverted commas), with a few lines from *Poems of Felicity* he fancies or dislikes, and stop when he feels tired.

The most inexperienced student should try to make each essay or paper a considered piece of decent craftsmanship; the weakest essay put together with some conscious care for content, structure and style will be better than a muddle.

Every essay has to be selective; it is impossible to put down everything conceivably relevant to the subject. So the student should first sort out what matter he wants to put into his essay. At this stage some jottings are useful; they may not be very orderly, just headings of the main topics or ideas to be mentioned. The first job is to *delimit the content* of the essay. The interested, capable student usually finds there is far more to discuss than time could possibly allow. If the difficulty is not to define a reasonable scope for the current effort, but to think of anything to say, the student has not yet done enough reading and thinking.

Suppose the subject is, 'Consider the use Keats makes of colour, light and shade.' The wording of the title already defines the subject fairly narrowly; but the student has still to settle some questions about content. Shall he read again quickly through Keats's *Poetical Works* – perfectly possible if he reads on a level of attention not close enough for the full experience of poetry – noting all lines with colour, light or shade references? or shall he confine himself to a smaller field such as *Endymion* or the 1820 volume? If the former, he will need in his first paragraph some phrase such as, 'A general survey of Keats's poems provides . . .'; if the latter, something like, 'From a detailed examination of the 1820 volume, we find . . .'. The range should be defined. Then the student will have to decide how he will treat the matter he has collected. Clearly a mere list of colour, light and shade references will not do; at university level he is expected to consider what these details are for; are they frequent enough to amount to a characteristic imagery? are they mostly there to make word-pictures more vivid and beautiful? are they often associated with myths or traditions? or are many of them in some way metaphorical or symbolic?

On the other hand, if the student is just told, 'write on *Hyperion*', he knows which poem must have his closest attention, though he should not limit his reading to *Hyperion*; he needs to look at *The Fall of Hyperion* and decide if he wants to mention the relationship between the two poems; and unless he has read a fair selection of Keats's other poems he cannot see how far *Hyperion* is different in style or content. This time he has himself to decide what aspect or aspects he will write about; will he emphasize its place in Keats's development, or place it in the epic tradition, perhaps relating it to *Paradise Lost*? or consider Keats's treatment of Greek mythology and its philosophical implications? Or will he make a close technical study of the poem, examining details of the versification, choice of words, sentence structure, imagery and so forth? or find some other line of approach, possibly supporting or refuting some critical opinion?

When the student has more or less delimited the range of his essay, assembled most of the material he is likely to need and sorted out his own opinions, he should spend some time *planning the structure* of his essay. A good essay never resembles a dead jellyfish: it has bones. The content must be organized.

Not every successful student makes a full written plan before starting his essay; some find a plan in the head is enough. A plan has not been passed by the local authority; it is permissible to deviate from it if, during the writing, a better line of argument presents itself. But any student who is having trouble with essays, 'not knowing what to say', 'can't get started', regularly getting disquietingly low marks, and so on, should try making a written plan of each essay. (He may also take an essay that has been slated, and try to write down, from the essay, its essential plan. He may well find this is impossible; if so, he has learned a lot.)

Any plan should be based on two simple principles:

1 an essay should have a beginning, a middle and an end;

2 the middle should follow, in a sequence of linked paragraphs, from the beginning, and lead into the end.

One possible structure is to begin with a generalization, then amass evidence to support it, and at the end come back to the generalization, preferably adding something: a new twist, a reservation, a pointed correction of the wording. Or an essay could begin with a question: 'Are Dryden's heroic figures no more than mouthpieces for elegant rhetoric?', 'Does Mrs Gaskell reduce social problems to sentimental melodrama?'; examine the question, marshalling some examples in evidence, and conclude by answering the question: 'It is clear from this examination of *All for Love* and *Don Sebastian* that Dryden, within his framework of unrealistic conventions, was quite capable of differentiating characters and varying their speech styles.' 'Thus we see that in *North and South* a relatively small element of melodrama in the plot helps to hold the reader's interest, but the general treatment of social problems is realistic and serious' – or whatever our opinions may be.

A first sentence may quote some critic, or some sentence or phrase from the primary work, using it as a kind of motto; the concluding sentence could repeat this, modify it, perhaps counter or reinforce it with another quotation from the same author. Suppose, for instance, that an essay must be written on Donne, and the student wishes to treat of Donne's reaction against Petrarchan love conventions that had worn out, Donne's general tone of intellectual vigour and emotional intensity; then the essay could begin with a quotation from Lucius Carie's *Elegy on Dr Donne*[2]

> he was a two-fold Priest; in youth,
> Apollo's; afterwards, the voice of Truth. . . .

and, after a study of some of Donne's love poetry and religious poetry, conclude, 'Thus it may be said that whether he wrote as Apollo's priest or an Anglican priest, Donne generally spoke with a voice we hear as essentially the voice of Truth.'

Good essays will not be written just by imitating one or other of the structures suggested above; there must be dozens of possible structures; the important thing is to give the essay some structure. Nor, of course, are my specimen conclusions necessarily true; but every essay should have its (provisional) conclusion.

It is important that the paragraphs in the solid middle of the essay should follow one another in a sequence of argument or exposition. The writer can to some extent check this by trying to link the first sentence of a new paragraph with the last sentence of the previous one:

> are examples of Keats's relating images of light to love and joy.
> This use of light may be compared with such images of darkness as . . . .

Words such as *therefore*, *so*, *thus*, *on the contrary*, *however*, *in contrast*, *nevertheless*, *for example*, *moreover*, often help to fasten these links. It will not always be possible to provide instantly obvious links to paragraphs; sometimes the student is introducing a completely new aspect of the subject; he may even be aiming at a startling shift of viewpoint; but, whatever the nature of his sequence, he should know what he is doing and why, not just ramble on.

Style is not as important as accurate content and a meaningful sequence of ideas, but it does matter. Waffle, padding and tangled sentences suggest a woolly mind and waste everyone's time; careless wording may make us say what we did not intend, or commit us to absurdly sweeping statements; a meagre vocabulary may force us into awkward repetitions or unwitting imprecision; even bad spelling can sometimes make us say what we did not intend ('this is my weakly essay . . .'), or incline the reader to suspect our reliability in more important matters.

Few of us are ever going to achieve a really individual, distinguished style such as gives a reader positive, conscious enjoyment; this is a difficult and rare achievement. If prose stylists of the calibre of Addison, Johnson, Lamb, Jane Austen, Macaulay, Carlyle, Dickens, were found in every street, we should hardly have our present concept of 'literature'. Some writers of detective stories and thrillers write far better prose, with respect to clarity, apt choice of words, economy and wit than most of us can ever hope to write. The student of literature who is aiming at serious authorship will of course be training himself all the time; most students may legitimately be content to aim at, merely: clarity, conciseness, correctness and consistency.

Did I say *merely*? Yes; these are minimum essentials for any good written work. Yes; but achieving this minimum costs most of us some painful effort.

It is little use having bright ideas if we cannot make other people understand them; indeed, if we cannot express them clearly we have probably not thought them out fully. The essay writer should keep in mind an imaginary reader – considerably less intelligent than a university teacher; and aim at making every fact, every stage in argument, crystal clear to that person. It may also be useful to picture the appropriate tutor asking those awkward 'what-do-you-mean-by' questions that are part of his duty! We should re-read whatever we have written, asking ourselves, 'Have I put this plainly?', 'Can someone else follow this immediately?', 'Am I sure this has only one possible meaning?' If in doubt, we can often learn something from reading the sentence or paragraph aloud, with an imaginery (or even a real) listener. If, apart from any rare word such as *chthonic* or *metempsychosis*, it cannot be read aloud with easy flow, there is probably something wrong with it.

Conciseness is a courtesy to the reader or listener, and especially the marker. Waffle makes large demands on people's time and gives hardly anything in return. An introduction should not include matter from which nothing eventually follows. Statements unsupported by evidence such as quotations are insufficient; but for many purposes three illustrations are as good as six; and quotations are no more a substitute for your own ideas and arguments, than arguments are any substitute for evidence.

Précis writing is a useful training in economy of words; so is the habit of taking real notes, as opposed to indiscriminate copying. An adequate working vocabulary is a help:

> Gibbon, more than many writers, was in the habit of using sentences that put in a number of explanations or other extras first, and then come to the most important statement at the end.
>
> *Frequent* use of *periodic* sentences is a *characteristic* of Gibbon's style.

A student of English, of all subjects, should respect his own language and write it correctly. Opinions on some details vary, but it is still true that every sentence should have a subject and finite verb; if it does not, it is probably not clear. We need only 'check' a reference, not 'check up on' it; redundant prepositions of this type waste time, dilute useful verbs and enfeeble our prose rhythms. *As* and *like*, *less* and *fewer*, *disinterested* and *uninterested*, are not interchangeable;

commas and semi-colons have specific functions; there are exact rules for the sequence of tenses, so that 'If we study Henry King's lesser-known poems, we should find . . .' is incorrect; also we should not muddle our pronouns as in 'if you look carefully at Hopkins's experimental metres, we shall find that . . .'. *Definitely* is not just a pompous synonym for *very*. There is no room here for the hundreds of points I would like to make, but the student whose work becomes measled with red for stylistic faults should spend some time with a modern work on English usage and perhaps with an old-fashioned grammar.

Finally, the style of an essay should be consistent, remaining within one stylistic register, unless, rarely, for some calculated special effect. Any essay that keeps shifting from, say, Johnsonese (probably not as lucid as Johnson's) to trendily colloquial, or from an abstract analysis full of conjunctions to an emotional impressionism full of adjectives, is likely to be confused and perhaps to have unintended comic effect.

The commonest cause of absurd stylistic incongruities in student essays is literary shoplifting. Either the student borrows a fine passage from a good critic and this exemplary style shows up his own clumsiness; or he borrows a piece of lumpy jargon that to him sounds clever, and shows up both the inferior critic and himself by putting two kinds of unformed style in juxtaposition. Francis Quarles has excellent advice here:[3]

> If thou intend thy Writings for the publique view, lard them not too much with the choice Lines of another Author, lest thou lose thy owne Gravy: What thou hast read and digested being delivered in thy owne Stile becomes thine: It is more decent to weare a plaine suit of one entyre cloth, than a gaudy garment chequer'd with divers richer fragments.

Every quotation should be placed in inverted commas, and

acknowledged either in advance – 'As Leavis says in *The Great Tradition* . . .'; or afterwards – '. . . F. R. Leavis, *The Great Tradition*'. A small alteration made to fit a quotation into your own sentence structure, say 'he is' for 'I am', should be indicated by square brackets. Though we do not put a paraphrased statement in inverted commas, since it is no longer a direct quotation, we must still acknowledge where it came from. Quotation, which modestly admits that we could not put it so well, is entirely legitimate; transcription palmed off as original composition is dishonesty.

17

EXAMINATIONS

Scribble thou, scribble thou, rail or write,
Write what thou wilt, I shall thee requite!
John Skelton, *Poems against Garnesche*[1]

the Tormentors examined him all the while, and to no Purpose, since he could not well confess that of which he was utterly ignorant.
Joseph Morgan, *A Complete History of Algiers*[2]

Any graduate who feels that three years' study of English Literature have given him nothing but the BA that to him stands for Bread Admission has largely wasted his time. He should have had also enormous intellectual development and delight, and considerable emotional enrichment, in an experience that may go on working in him for decades. However, most of us need bread tickets; most of us, frankly, need some fairly crude goals, tests and incentives to reinforce our purer motives; and a degree certificate, with its class, is a tolerably objective proof that someone has worked through a systematic course, and how conscientiously. A good degree is often some economic advantage and a personal encouragement; it is only common sense to aim at doing full justice to oneself.

The main factors in examination success are actual organized knowledge, and skill in applying it to the examination situation.

The time to start building up the stock of organized knowledge is the first day of the university course. The final examination is designed to test three or more years' work, and nobody can do three years' work in three months. The student who has a spell of sheer laziness somewhere in the course, or a period of real troubles that handicap him for work, must not despair; a patch of very hard work can usually repair much of the damage. Most of us throughout life manage to climb a few ladders after slithering down a few snakes; but climbing ladders in haste takes more effort than walking on a level. At almost any stage, a spell of exceptional and sustained effort is likely at least to mitigate disgrace; but panic cramming is unhealthy and largely disagreeable, whereas the work properly distributed should include much joy; and cramming is far less effective than steady moderate effort.

Examination technique is not esoteric. It is largely common sense.

I recently calculated that in my life I must have marked about seven thousand examination papers on English; and I am certain how most examinees who seem to have done some conscientious studying lose most of their marks, sometimes wrecking whole answers. They do not reply to the question asked.

What happens is probably something like this. The candidate sits down, with a bellyful of butterflies, and stares at the question paper as if it were a cobra; sees a question on an author on whom he has worked hard:

> *7. How far do Tennyson's dramatic monologues succeed as presentations of character?*

Ah! Tennyson, yes, I can do something on Tennyson . . . Tennyson's presentations of character . . . now . . . ah, King Arthur, Elaine, Gareth, Vivian . . . Enoch Arden, with Philip, and Annie; oh, and the way the personality of

Arthur Hallam shows through in *In Memoriam* . . . only an hour to get it all down, quick, don't waste time, start with Enoch. . . .

And the candidate, who really does know something about Tennyson, writes a badly constructed, but in a scrappy way fairly perceptive answer about some of Tennyson's characters, even with a few nearly accurate quotations; fifty-three minutes gone, a third of the paper. . . . *and earns no marks at all*, because he has not mentioned a single dramatic monologue. *Maud* was the obvious essential, with plenty of other dramatic monologues to choose from, such as *Oenone*, *St Simeon Stylites*, *The Grandmother*, *Northern Farmer*, *Rizpah*, *Columbus*, *The Spinster's Sweet-Arts* and many more poems in which Tennyson imaginatively impersonates someone else.

It seems hard that honest work and good intentions should be wasted; but the cruel examiner is not to blame; the fault lies with the muddled candidate. It is sad, but it is not unjust.

Every now and then someone waxes eloquent about the essential unfairness of examinations; years of work are tested in an ordeal of a few days; students cannot do justice to themselves under stress and in such an artificial situation.

Of course examinations are unfair. They are set and sat by fallible creatures in an imperfect world. Any test that has to be staged on specific dates has an element of unfairness; about 20 per cent of the women will inevitably not be at their best, plus some percentage of both sexes who have ailments, injuries, or some distracting misfortune such as being bereaved or jilted. (Still, I do recollect an Indian student who obtained his degree when poverty had reduced him to sleeping on a railway platform and beriberi was sapping his vitality. . . .)

However, university examinations may well be among the least unfair tests in a world full of unfairness. The average student would be amazed if he could be a fly on the wall at the examiners' meetings for setting papers and awarding final marks: he would see how much trouble is taken to make both question papers and marking as fair as possible.

Solving problems in a limited time is not some exceptional torture inflicted by sadistic professors on students. It is the demand constantly made on most of us in normal life. What about the student who catches a lecturer bookless in a corridor and asks a searching question? Anyone whose work is not wholly repetitive is daily applying the appropriate portion of specialist knowledge to a series of specific problems: the vicar, doctor, lawyer, nurse, plumber, policeman, gardener, farmer, sailor, soldier, secretary. I expect five friends to supper; ask the butcher for six large lamb chops; if, on unwrapping my purchase, I find one chop, half a pound of liver, and a lot of tripe, I shall not be pleased. A mass of knowledge from which we cannot quickly select what we need for a particular purpose is of little use to us.

Solving problems under stress is also an everyday requirement of normal life. When we have to decide what to do about an unreasonable parent, a naughty child, a marital quarrel, a friend in distress, difficulties with colleagues, money troubles, overwhelming temptations, conflicting duties, an accident or emergency, life does not obligingly stand still and quiet while we sort out our ideas and come to our well-considered conclusion. At best, all life's other demands continue pressing on us; at worst, other people may deliberately put pressures on us with conscious intent to distort our judgment. Somehow we have to draw on what little understanding and wisdom we have, when circumstances make cool judgment very difficult.

In contrast, everyone is trying to divest the examination of all avoidable stress. Candidates are provided with a quiet room, shielded from interruptions, noise and other people's untimely demands; officials organize soothing fixed routines, known procedures, convenient amenities; if a candidate is visibly in distress, an invigilator will try to encourage, comfort and calm him; if he is ill or handicapped, arrangements will be made to help him.

So, though we can make jokes about our inquisitors, we should not flop into exaggerated self-pity or childish resentment about examinations. The student who has real, crippling examination phobia is a very exceptional case indeed, is ill, and should seek professional advice as early as possible. The rest of us just have 'butterflies', 'nerves', 'funk' or 'collywobbles'; it is perfectly normal to feel nervous before a minor ordeal, and, as for the orator or actor, a preliminary fit of 'nerves' is usually a *good* sign; the mind is, as it were, revving up to full speed.

Some students have some real difficulty in putting down their organized ideas at the speed required. The time to cope with this is as long as possible before Finals, for the remedy is practice. The student for whom this is a serious problem may be well advised to write most of his weekly or other essays to a time limit; this should be discussed with a tutor, for no one can both write to time and be verifying all his references. A tutor may be willing to look at some extra exercises, perhaps questions done to time from previous Finals papers. And termly or occasional tests are useful, not only to check that students are doing enough work, but also to keep them in practice for examinations.

I repeat that the long-term preparation for relatively easy success in Finals is three years of steady work, keeping up with each week's and each term's programme.

The final term should include a fair amount of revision, on which tutors can give some advice. Some aspects of revision are: re-read, probably several times, all the three years' notes and essays, not mechanically, but with self-criticisms and corrections; memorize facts likely to be needed and not already fixed in the mind; re-read important books that were studied some time ago; be sure that knowledge of set texts is detailed enough and is well consolidated; look at some question papers from previous years – not to attempt 'question-spotting', for this year, which is as dangerously fallible as a newspaper horoscope, but to see what *kind* of question will be asked.

Anyone may be unlucky; but, as far as it lies within your own control, keep fit. As Octavius advised Cleopatra, 'Feed, and sleep:'[3] Regular, dietetically sound meals, enough sleep, daily fresh air and exercise, do help. It is a mistake to become dependent on cigarettes; not only are they unhealthful and expensive, but they are forbidden in examination rooms, and no student needs withdrawal symptoms from a minor addiction to supplement examination nerves. Heavy or habitual drinking, even serious overeating, impair mental work and slow down reactions that, for writing to time, should be as quick as possible.

The startling effectiveness of many modern medicines has led some of us almost back to a pathetic medieval faith in magic pills. Never, never take any drug affecting the brain – pep pill, tranquillizer, trendy illegal treat – when it could possibly affect an examination performance, unless it is something on a current prescription from a doctor who knows the date of your examination as well as your medical history. Most experienced examiners have seen the pitiful papers, with two lines to each answer, total irrelevance, delirious incoherence, of candidates betrayed by drugs.

A less catastrophic, but perhaps harmful, mistake is to cram until the last minute. It may be possible to memorize a useful date, an elusive quotation, a couple of spellings, the night before Finals. To try to learn any substantial new material at this stage may be worse than a waste of time; we do not easily recall what we have only just learned, and may garble it, over-emphasize it in an answer, or confuse some other knowledge that was previously in good order. The best preparation the night before Finals is a spell of mild and safe pleasure such as an enjoyable film or television programme, or an evening with friends; an easy walk and an early night, first making sure before going to bed that everything is ready for morning: clothes, shoes, a writing instrument *with at least one spare*, spectacles if needed, ruler, handkerchief, instructions where to go and anything else likely to be wanted. A fuss in the morning – 'can't find my pen!' – 'shoelace gone – I knew it was thin!' – 'is it Ascham Hall or the Mulcaster Building?' – is the last thing a nervous candidate wants. I offer one other hint from experience; we have no right to risk vexing other candidates by noisy sucking or distracting smells, but something unobtrusive such as a glucose tablet or two taken about two-thirds of the way through the paper may give a useful emergency lift.

A candidate who has worked reasonably well must now put some trust in his unconscious mind. There can be a hideous moment outside the examination room when the student who is about to earn a First feels he knows nothing. But memory recalls things in response to some stimulus, not in a non-situation; when the student sees a question, the necessary knowledge will come flooding out – provided that he previously put it into his memory.

The first thing to do when confronted with a question paper is to *read the rubric*. An instruction to 'answer one question from Section A, one question from Section B and two other questions' must be obeyed. Answer two questions from Section A and two from Section C, and 25 per cent of the possible marks have gone already, even if there is not a single mistake in the answers.

The second stage is to *read the questions*, with a view to choosing the most congenial ones: you know more about Herbert than Crashaw, you have *Samson Agonistes* nearly by heart but never really got into *Comus*.

The third stage is to *study carefully the exact words* of the first question you wish to answer, before writing anything more than your name and the number of the question. Never mind if everyone else is already writing.

It is better to spend even ten minutes working out what is required to answer the question set, and forty minutes writing an answer all of which is to the point, than to spend sixty minutes answering a question that is not there.

Many students find it helps to jot down a few headings as a rough plan; there is not time to do a detailed outline; but have some idea where you are going, before you start.

Time must be allocated sensibly. University marking varies in flexibility. You may even be told, 'Answer two, three or four questions', in which case two can be done in more depth, three or four in less depth; but if, for instance, there are three questions to be done in three hours, about fifty minutes should be allowed for each question. If you run out of time on the last question, dash down as lucid an outline in note form as you can; if it makes sense and is correct, you will earn at least some marks. Do not count on one brilliant answer to compensate for a weak one or no answer to the last question at all; it is difficult to gain more marks on the former than you lose on the latter.

Try to write legibly. A scrawler will not reform on the first morning of Finals; but a student whose writing is a nuisance to everyone else should try earlier in the course to improve it. A paper that is easy to read puts the examiner in a favourable frame of mind. He will not, in theory, deduct marks for those beetle-tracks that strain his eyes and waste his time. What he may do is refuse to give you the benefit of any doubt or show any sympathy over a misunderstanding. And papers can be officially 'declared illegible', which may cause the candidate a great deal of expense and inconvenience. Book titles should be underlined, as should the titles of single poems or essays; it is correct practice; it makes reading easier; and sometimes it does matter whether you are referring to Hamlet or Hamlet.

Accuracy is obviously important; the examination is

testing knowledge as well as judgment. There is no need to break your heart if you realize at night that you made one wild hilarious howler. Examiners know that nervous students, writing in haste, can make ludicrous slips of the pen, such as attributing *Hydrophobia* to Sir Thomas Browne. (Really funny howlers are often passed round and improve rather than fray tempers!) A constant slovenliness makes a far worse impression; one bad mistake may be an accident; a paper on which thirty things are not-quite-right suggests three years of careless work. To copy something, even spelling, from the examination paper incorrectly looks very bad: can this candidate read *nothing* carefully?

The basic technique of an answer on literature is almost always to argue a case, agreeing or disagreeing with a statement in the question, or looking at both sides. It is important to notice the exact wording of the question and carry out the instruction: 'Comment', 'Justify this statement', 'Justify or refute this statement', 'Discuss' (which requires at least two possible points of view); 'Analyse', 'Compare', 'Explain', 'Elucidate', 'Illustrate' (which requires a good stock of examples proving some point), 'Trace'. If you are asked 'Do you agree?' you may answer either 'Yes' or 'No' or even something midway, such as 'In general this assertion is true, but there are exceptions. . .'. It is always worth while to pause and consider exactly what shape of answer corresponds to the wording of the question.

Every assertion made in arguing a case should be supported by evidence. University examiners are not interested in whether you agree with them or are taking some correct line; they want to know whether you have studied the book and can argue about it honestly, intelligently, relevantly, cogently and with reasonable evidence.

Quotations usually provide much of the evidence. Every quotation should be in inverted commas; and verse should be set out in its lines. If you are not sure of the exact words of a verse quotation, at least check that it scans. Shakespeare wrote: 'Here is my throne, bid kings come bow to it.'[4] The forgetful student who puts 'bid princes bow to it' or 'bid kings come kneel to it' does better than the one who writes: 'Here is my throne, command kings to come and bow.' Shakespeare might conceivably have written the first two; no one with an ear for blank verse could perpetrate the third. A paraphrase or a recognizable allusion is often acceptable instead of a quotation, unless the question demands exact words illustrating stylistic detail.

The last few minutes of each examination should be reserved for *re-reading*. The candidate who looks through his paper may correct several slips (even a disastrously omitted 'not'!) and perhaps add one or two useful points.

Once he has left the examination hall, however, the student should say, 'What I have written, I have written.' Now it is the next paper that matters. Students who go into huddles for post-mortems waste energy and frighten themselves.

Soon, we hope, comes the last episode of undergraduate life, the usually beautiful degree ceremony with its grand moment of legitimate pride.

And that makes a very good . . .

beginning.

18

SOME USEFUL BOOKS

A comprehensive reading list for the first-degree student of English Literature would be bigger than this book. The present list merely suggests a few books that may be helpful to a student either before embarking on a first-degree course, or early in the course. (The publishers are all London publishers unless another place is specified.)

EFFICIENT STUDY

One of the best of several sensible books available is Harry Maddox, *How to Study*, Pan Books, 1963 or David & Charles, Newton Abbot, 1970. Thorough and scientific, but quite easy to read; particularly helpful on 'Learning and Remembering', 'Notes and Lectures' and 'Thinking'. A good alternative is D. E. James, *A Student's Guide to Efficient Study*, Pergamon Press, Oxford, 1967, which has useful detail on such subjects as preparing a paper for a group.

Reasonably good books, sound and helpful but not quite so meaty, include Hunter Diack, *101 Aids to Exam Success*, Dickens Press 1967; *Study, the Easy Way*, Transworld Press, 1967. (Do not be put off by the slick titles; Diack has published more specialist works on educational psychology.)

Derek Rowntree, *Learn How to Study*, Macdonald, 1970. (A programmed textbook, which may help some students to read it attentively.)

Colin E. Woodley, *How to Study*, Angus & Robertson, Sydney, London, etc., 1959.

A good book on efficient reading for the student is Owen Webster, *Read Well and Remember*, Hutchinson, 1965. And for fun, try Plutarch, *On the Student at Lectures*, written probably between 70 and 120 AD, to be found in Plutarch, *Selected Essays*, trans. T. G. Tucker, Oxford University Press, 1913. Some of it is now entertaining if read aloud with due solemnity, but there is much permanently sound advice on the art of being educated. Students may have changed in eighteen centuries less than we might expect. . . .

COMPETENT WRITING

We can all go on trying, to the end of our days, to improve our own writing of our magnificently supple, subtle and multifarious language; and there must be some scores of relevant guidebooks, as well as, more interesting, the examples given by great writers. For mending our most obvious faults, useful books include:

Spelling: Hunter Diack, *Spelling, the Easy Way*, Transworld Publishers, 1967.

Punctuation: Eric Partridge, *You have a Point There*, Hamish Hamilton, 1953; Routledge & Kegan Paul, 1978.

Usage: Eric Partridge, *Usage and Abusage*, Hamish Hamilton, 1947. H. W. Fowler, *Modern English Usage*, second edn, revised by Sir Ernest Gowers, Oxford University Press, 1965.

Vocabulary: Obviously, any good dictionary. Also Peter Roget, *Thesaurus of English Words and Phrases*, revised by D. C. Browning, Dent, 1966.

General: Eric Partridge, *English, a Course for Human Beings*, Macdonald, 1949. Intended primarily as an English course for school pupils from ten to eighteen; but the undergraduate who has become aware of holes in his school training may find material to mend them here; and Partridge, maintaining high standards, keeps a friendly, agreeable tone. This course includes, besides excellent expositions of grammar, punctuation and composition, useful lucid chapters on 'Rhetorical Terms', 'Metric or Prosody' and 'Appreciation and *Lecture Expliquée*'.

CONCEPTS OF CRITICISM

R. L. Brett, *An Introduction to English Studies*, Arnold, 1976. Slight and scanty as compared with the other books in this section, but sensible and useful for the beginner.

Helen Gardner, *The Business of Criticism*, Oxford University Press, 1959. (The first three essays, under the heading 'The Profession of a Critic', should help any student of literature.)

F. R. Leavis, *English Literature in Our Time and the University*, Chatto & Windus, 1969, raises many questions that the serious student should consider.

C. S. Lewis, *An Experiment in Criticism*, Cambridge University Press, 1961, seems to me a quite exceptionally wise, sane, honest book. It is very easy to read; to live up to it is not so easy.

Angus Ross (ed.), *English, an Outline for the Intending Student*, Routledge & Kegan Paul, 1971. Eight useful essays, of which the most generally applicable are probably those by G. K. Hunter, D. J. Palmer, G. Josipovici and L. Lerner; with some practical information.

CRITICAL VOCABULARY

Probably the most ample and useful, as well as the most up-to-date, is J. A. Cuddon, *A Dictionary of Literary Terms*, Deutsch, 1977, which includes many items not easy to find elsewhere, such as *happening*, *pruning poem*, *serpentine verse*, *shaggy-dog story*. . . . Eminently worth owning; but rather expensive. A good second choice would be K. Beckson and A. Ganz, *A Reader's Guide to Literary Terms*, Thames & Hudson, 1961.

For modern critical developments (and jargon?) try Roger Fowler, *A Dictionary of Modern Critical Terms*, Routledge & Kegan Paul, 1973, which does much to help the reader with numerous semantic difficulties.

On poetry in particular: George Saintsbury, *Historical Manual of English Prosody*, Macmillan, 1910, does not go beyond Swinburne, but is still a very useful, though heavy, handbook of traditional verse forms.

Alex Preminger (ed.), *The Princeton Encyclopaedia of Poetry and Poetics*, Princeton University Press, 1965, enlarged edition 1974, is a magnificent reference book; its 992 double-column pages contain a great deal more than the average student is ever likely to want, but a great deal that he needs. Any student could profit from such articles as 'Allegory', 'Ballad', 'Catharsis', 'Courtly Love', 'Irony', 'Ode', 'Rhyme', 'Satire', 'Verse and Prose'.

HISTORY OF LITERATURE

Every student needs to possess a one-volume history of literature, both to read for a perspective and to consult for checking. Suitable ones for work at university level, sound, solid and thorough, are: George Sampson, *The Concise*

Cambridge History of English Literature, Cambridge University Press, 1941; E. Legouis, L. Cazamian and R. Las Vergnas, trans. H. D. Irvine, *A History of English Literature*, Dent, 1964.

As extras: Peter Quennell and Hamish Johnson, *A History of English Literature*, Weidenfeld & Nicolson, 1974, is a beautiful illustrated history, anecdotal and popularized; it is not suitable as a reference book for the university student, but someone who has felt damped down by set books and minutiae might well find it refreshing, recover a sense of grand vistas and vividly alive authors, and catch fire again.

J. B. Priestley, *Literature and Western Man*, Heinemann, 1960, has Priestley's usual gusto, good sense and genuineness, and very usefully places English Literature in a larger framework, relating it to other European literatures.

Useful and affordable paperback reference books include:
Boris Ford (ed.), *The Pelican Guide to English Literature*, Penguin, Harmondsworth, seven volumes (various dates with various revisions) of intelligent essays and useful bibliographies.

David Daiches (ed.), *The Penguin Companion to Literature*, vol. 1, *British and Commonwealth Literature*, Penguin, Harmondsworth, 1971, mostly a dictionary of literary biography. The other three volumes, embracing the literature of the rest of the world, are also well worth having.

Other works of reference:
J. C. Ghosh and E. G. Withycombe, *Annals of English Literature 1475–1925*, Oxford University Press, 1936, for quick checking of chronological relationships.

Sir Paul Harvey, *The Oxford Companion to English Literature*, Oxford University Press, 3rd end, 1946. Includes, besides biographies of authors and brief accounts of many literary works, identifications of many characters, mythological allusions, many important foreign works, and miscellaneous points about, for example, Banbury, Corn Laws, Elgin Marbles, Limbo, Pandects or Zimbabwe.

BACKGROUND KNOWLEDGE

Theatre History

Phyllis Hartnoll, *The Oxford Companion to the Theatre*, Oxford University Press, 1951, includes, besides biographies of dramatists and performers, many articles on dramatic genres and the history of theatrical techniques, e.g. under 'Melodrama', 'England', 'Stage'.

Allardyce Nicoll, *The Development of the Theatre*, Harrap, 1927, a notably clear and well illustrated account of the history of the stage itself, as opposed to the drama.

Greece and Rome

Probably the best all-round handbook, covering biography, antiquities, mythology, history and literature, and especially helpful on actual literary works, is:
Sir Paul Harvey, *The Oxford Companion to Classical Literature*, Oxford University Press, 1937 (latest corrected reprint, 1974). The student who can afford to own this should do so.

Very useful, but not so comprehensive, is:
John Warrington, *Everyman's Classical Dictionary*, Dent, revised edn, 1969.

Much better than nothing, but confined mostly to literary biography:
D. R. Dudley and D. M. Lang, *The Penguin Companion to Literature*, vol. 4. *Classical and Byzantine*, Penguin, Harmondsworth, 1969.

Also useful:
H. J. Rose, *Outlines of Classical Literature for Students of English*, Methuen, 1959.

J. W. H. Atkins, *Literary Criticism in Antiquity*, 2 vols, Methuen, 1952, includes more than most students of English Literature will want; but even a quick reading may be an eye-opener as to how far literary criticism had already developed in ancient times.

F. A. Wright and T. A. Sinclair, *A History of Later Latin Literature*, Routledge, 1931. Treats of Latin Literature from the fourth century AD to the middle of the seventeenth century, a subject often of interest to the student of English Literature, on which information is not as easily found as on classical Latin.

Mythology
Classical dictionaries identify most of the classical mythological allusions the student needs.

A more comprehensive volume is:
Larousse Encyclopaedia of Mythology, ed. Felix Guirand, trans. Richard Aldington and Delano Ames, Batchworth Press, 1959, a survey of world mythology, with plentiful illustrations; very helpful on Greece and Rome; the Celtic mythology is useful for e.g. Yeats studies, the Teutonic for e.g. William Morris; the whole volume is fascinating.

E. C. Brewer, *Brewer's Dictionary of Phrase and Fable*, Cassell, 8th revised edn, 1963 (first published 1870), is an inexhaustibly interesting treasury of mythology, folklore, proverb, idiom, symbols and assorted allusions; many books in one, it is well worth owning if possible.

Medieval Background
Obviously, any reputable historical work treating of the

medieval period will contribute something to understanding; but a very valuable extra, specifically for the literary student, is:
C. S. Lewis, *The Discarded Image*, Cambridge University Press, 1967.

Religious Background
Anyone writing on this subject is likely to hold some fairly firm belief or disbelief, and a totally 'unbiased' account may well not exist.

Bamber Gascoigne, *The Christians*, Cape, 1977, readable as a novel, very well illustrated and trying to be objective, is a sketchy, anecdotal, occasionally flippant popularization, but should give the student who has no background knowledge a useful start, and some notion of the range and fascination of the subject.

Life Magazine editorial staff, *The World's Great Religions*, Collins, 1959. The sections on Judaism and Christianity of this beautiful illustrated volume may provide some helpful start for the student who knows nothing of Christian background.

James Hastings (ed.), *Encyclopaedia of Religion and Ethics*, 13 volumes, Clark, Edinburgh, 1908–26, is the great standard reference work covering the world, a mine of information, astonishment and interest; the student of literature will find valuable help in such articles as 'Western Church', 'Protestantism', 'Symbolism (Christian)'.

R. A. Knox, *The Belief of Catholics*, Benn, 1927, was written with intent to convert non-Catholics, not to inform literary students, but is useful for the latter purpose; note its date, well before the epoch-making Papacy of John XXIII.

The student who has not read most of the Bible has already

handicapped himself severely for understanding English literature.

Lastly, two possible firelighters:
John Livingston Lowes, *The Road to Xanadu*, Constable, 1927; because even for the beginner this study of how Coleridge used raw materials from his reading to create his two greatest poems, with 'the imagination voyaging through chaos and reducing it to clarity and order', is one of those critical works that communicates both the marvellousness of the creative mind, and the excitement of scholarship.

F. W. Bateson, *The Scholar-Critic*, Routledge & Kegan Paul, 1972. Intended primarily for students beginning post-graduate work, this book may be a trifle daunting to the novice; but the novice who can face it will find much intellectual stimulus, and a quietly noble presentation of the ideals and methods of real scholarship.

NOTES

Chapter 1 Why study literature?

1 G. K. Chesterton, *Charles Dickens*, Methuen, London, 1906, p. 99.
2 'They are there, a hundred cubits tall, Christ at the head, Homer in the midst, all the warriors of ideas, all the gladiators of God.' Victor Hugo, *Les Contemplations*, Le livre de poche edn, Paris, 1968, p. 454.
3 Rebecca West, *The Meaning of Treason*, 1949, Reprint Society edn, 1952, London, p. 19.
4 *King Lear*, III. vi. 82–3. (All line references to Shakespeare's plays are taken from the one-volume Oxford University Press edition ed. by W. J. Craig, 1935.)
5 Rebecca West, op. cit., p. 65.
6 D. J. Enright, *The Terrible Shears*, Chatto & Windus, London, 1973; 'Escapism', p. 34.
7 Alexander Pope, *Essay on Criticism*, l. 298.
8 *Poems of John Clare*, ed. A. Symons, Oxford University Press, 1908, p. 89.
9 D. H. Lawrence, *The Rainbow*, Phoenix edn, Heinemann, London, 1955, p. 307.
10 Colin Dexter, *Last Seen Wearing*, Macmillan, London, 1976; Pan Books edn, London, 1977, p. 81.
11 Charles Dickens *Little Dorrit*, ch. 10, New Oxford Illustrated Dickens edn, Oxford University Press, 1953, p. 110.
12 John Wain, *Professing Poetry*, Macmillan, London, 1977, p. 271.
13 *King Lear*, IV. vii. 72–6.
14 *Pearl*, ed. E. V. Gordon, Oxford University Press, 1953, pp. 16–17; ll. 445–52.
15 *Troilus and Criseyde*, V, stanzas 260–1.
16 Anthony Trollope, *Phineas Finn*, World's Classics edn, Oxford University Press, 1969, pp. 269–71.

Chapter 2 Flying solo

1 Arnold Bennett, *Clayhanger*, Methuen Uniform edn, London, 1947, p. 85.

Chapter 3 Is the syllabus silly?

1 James Hulme Canfield, *The College Student and His Problems*, Macmillan, New York and London, 1902, p. 55. (J. H. Canfield was an American university president.)
2 M. F. K. Fisher, *The Art of Eating*, Faber & Faber, London, 1963, p. 464.
3 Alexander Pope, *Essay on Criticism*, l. 232.
4 Let me pay tribute at least in a note: the lecturer was Dr Catherine Ing, at Oxford.
5 Thomas Bowdler's preface to his *Family Shakespeare*, 1818.

Chapter 4 Do we murder to dissect?

1 From notes, 'A Vision of the Last Judgment', in the 'Rossetti MS.' Geoffrey Keynes's edn, *Poetry and Prose of William Blake*, Nonesuch Press, London, 1948, pp. 649–50.
2 *The Tables Turned* (Wordsworth, *Poetical Works*, ed. E. de Selincourt, Oxford University Press, 1940, vol. iv, p. 57.
3 Pope, *The Dunciad*, IV. ll. 211–12.
4 *Johnson on Shakespeare*, ed. Walter Raleigh, Oxford University Press, 1908, pp. 58–9.
5 Ibid., pp. 61–2.
6 *The Arte of English Poesie*, III. xii. Ed. G. D. Willcock and A. Walker, Cambridge University Press, 1936, p. 170.
7 *Preface* to Shakespeare, *Johnson on Shakespeare*, ed. cit., pp. 61–2.
8 *Men and Women*, ed. Paul Turner, Oxford University Press, 1972, p. 107, note p. 332.
9 Ibid., p. 372.
10 Wordsworth, *The Prelude* (1805), I. 614.
11 In a reference to the Balearians, ch. 1, para. 13. Sir Thomas Browne, *Hydrotaphia*, ed. W. A. Greenhill, Macmillan, London, 1937, p. 15.
12 Seamus Heaney, 'Strange Fruit', in *North*, Faber & Faber, London, 1975, p. 39.

Chapter 5 Relevance and reverence

1 Henry David Thoreau, *Walden*, Everyman edn, Dent, London, 1908, p. 90.
2 *Everyman's Dictionary of Quotations*, p. 380. I have not succeeded in tracing this to a source.
3 Louis MacNeice, *Autumn Journal*, Faber & Faber, London, 1939.
4 G. Bernard Shaw, *Pygmalion*, Act V. *Complete Plays of Bernard Shaw*, Odhams Press, London, n.d., p. 749.
5 I met this anonymous cleric in a notebook of Charles Reade's, but so far have not identified him.
6 C. S. Lewis, *An Experiment in Criticism*, Cambridge University Press, 1961, p. 19.
7 Pope, *Essay on Criticism*, I, 179–80.

Chapter 6 Owning, borrowing, consulting

1 Chaucer, *Prologue* to *The Canterbury Tales*, ll. 93–6.

Chapter 7 Editions

1 *Antony and Cleopatra*, V. ii. 282–7.
2 Nevill Coghill, *The Canterbury Tales*, Penguin Books, Harmondsworth, 1951.
3 The question is discussed at length in L. B. Osborne's *The Life, Letters and Writings of John Hoskyns*, Yale University Press, New Haven, 1937, pp. 285–7.
4 Jonathan Swift, *Gulliver's Travels*, II. 1.
5 Loc. cit.
6 *The Prologue*, l. 626.
7 See e.g. *Complete Poetical Works*, ed. E. H. Coleridge, Oxford University Press, 1912, vol. 1, p. 193.
8 Henry James, *The Ambassadors*. See edn of S. P. Rosenbaum, Norton, New York, 1964, pp. 353–67.
9 Scott Fitzgerald, *Tender is the Night*. See edn of Malcolm Cowley, Grey Walls Press, London, 1953, pp. xii–xix.
10 Some previous legislation, beginning with the incorporation of the Company of Stationers in 1557, and including an important 'Act for the Encouragement of Learning' in 1709, had done something to protect literary property; but the 1842 Act was the first to guarantee to authors fairly substantial rights.

Chapter 8 The author and the critics

1 *Love's Labour's Lost*, I. i. 84–7.
2 A double translation may have moved some distance from the original; I do not know Estonian and have translated from the Esperanto version by Hilda Dresen in *Estona Soveta Poezio*, Eesti Raamat, Tallinn, 1977, p. 31.
3 Pope, *Moral Essays*, III. 1.

Chapter 9 Some ways of misjudging

1 George Eliot, *Middlemarch*, ch. 20.
2 Thomas Hardy, *Tess of the D'Urbervilles*, Phase the Fourth, section 25.
3 In the series of prefaces now usually known as *Lives of the Poets*.
4 One of the 'Miscellaneous Sonnets'; it begins 'While Anna's peers and early playmates tread . . .' Ed. cit. p. 273.
5 See *The Poetical Works*, ed. E. de Selincourt, Oxford University Press, vol. II, p. 493.
6 *Shakespeare in the Soviet Union*, ed. Roman Samarin, Alexander Nikolyukin, trans. Avril Pyman, Progress Publishers, Moscow, 1966.
7 Essay by Ivan Anisimov, in ibid., p. 142.
8 Interview with Alexander Ostuzhev, in ibid., p. 158, the 'political significance' being anti-racist and anti-colonialist.
9 Ibid., p. 163.
10 Essay by Nikolai Okhlopkov, in ibid., p. 202.
11 Claude Roy, *Jules Supervielle*, Pierre Seghers, Paris, 1949, p. 17 (my translation).
12 Arthur Koestler, *Arrival and Departure*, Cape, London, 1943.
13 Ibid., pp. 185–6.
14 Johnson, *Preface* to Shakespeare, in *Johnson on Shakespeare*, ed. Walter Raleigh, Oxford University Press, 1908, p. 16.
15 John Wain, *The Living World of Shakespeare*, Macmillan, London, 1964.
16 Ibid., p. 21.

Chapter 10 Some ways of misreading

1 R. Austin Freeman, *Mr. Polton Explains*, Hodder & Stoughton, London, 1940, p. 45.

2 See e.g. James Sutherland, *Defoe*, Methuen, London, 1937, pp. 84–6.
3 *A Midsummer Night's Dream*, V. i. 21–2.
4 See A. Norman Jeffares, *A Commentary on the Collected Poems of W. B. Yeats*, Macmillan, London, 1968, p. 502, and Curtis Bradford, 'Yeats's *Last Poems* Again', Dolmen Press Centenary Papers, 1965, VIII, p. 274.
5 *Julius Caesar*, II. ii. 123–5.
6 *Hamlet*, II. ii. 583–642.
7 *All's Well that Ends Well*, I. i. 91–117.
8 Ibid., II. iii.
9 *King Lear*, III. iii. 1–21.
10 *A Midsummer Night's Dream*, IV. viii. 206–26.
11 *Hamlet*, III. iii. 36–72.
12. *Othello*, I. iii. 389–410.
13 *The Monkes Tale*, ll. 287, 289–300. The protagonists are now known as Zenobia and Odaenethus.
14 Ibid., l. 308.
15 *Richard II*, III. ii. 54–7.
16 See e.g. *Henry IV, part 2*, III. i. 4–31; 66–79; IV. v. 182–218.
17 *Henry V*, IV. i. 311–14, and see the rest of the speech.
18 Marlowe, *Tamburlaine the Great*, II. vii. 18.
19 For an interesting little debate that treats the chronological ideas current in Marvell's day, see letters by Roger Sharrock and E. E. Duncan-Jones in *TLS*, 13 October, 1958, 5 December 1958, 16 January 1959.
20 Anonymous, as far as I know.
21 J. L. Motley, *The Rise of the Dutch Republic*, World's Classics edn, 1906, III, p. 548.
22 John Webster, *The White Devil*, II. i. 135–6.
23 'The Aziola', *Complete Poetical Works*, ed. T. Hutchinson, Oxford University Press, 1934, p. 642.
24 *Macbeth*, I. vii. 45.
25 Philip Hope-Wallace, in the *Guardian*, 2 September 1975.
26 Charlotte Brontë, *Villette*, ch. 30.
27 Thomas Hardy, *Jude the Obscure*, III. ix. Macmillan uniform edn, London, 1929, p. 222.
28 *The Poems of William Dunbar*, ed. W. M. Mackenzie, Porpoise Press, Edinburgh, 1932, p. 69.
29 *The Poems of John Donne*, ed. H. J. C. Grierson, Oxford University Press, 1933, p. 83 (Elegie IX).
30 H. Gardner, *The Elegies and the Songs and Sonnets*, Oxford

University Press, 1965, p. 150.
31 *Julius Caesar*, I. iii. 32.
32 *Paradise Lost*, ix. 499–500.
33 *The Tempest*, I. ii. 464–5.
34 Antony and Araminta Hippisley Coxe, *The Book of the Sausage*, Pan Books, London, 1978, p. 130. Confirmed by *OED*.
35 *Guardian*, 7 September 1976.
36 *The Prologue*, l. 74.
37 *OED*, Second Supplement.
38 Quoted by Helen Gardner, *Notes and Queries*, 23, 5–6, May–June 1976, p. 195.

Chapter 11 Figs, dates and reasons

1 Lewis Carroll, *Complete Works*, Nonesuch Press, London, 1940, p. 510.
2 Crabbe, *The Borough*, VII, 128–31.
3 Byron, *English Bards and Scotch Reviewers*, ll. 235–64.
4 Leslie A. Marchand, *Byron*, John Murray, London, 1957, vol. II, pp. 528, 543.
5 Ibid., vol. I, 309n.
6 Ibid., vol. II, pp. 544, 624n, 753.
7 Ibid., vol. I, p. 325.
8 Edmund Blunden, *Shelley*, Collins, London, 1946, pp. 122, 170, 194.
9 Ibid., pp. 147–50.
10 Ibid., pp. 167–8, 177, 195.
11 Ibid., pp. 181–2, 184–5.
12 Ibid., p. 186.
13 H. Ben-Israel, 'Carlyle and the French Revolution', *Historical Journal*, 1, 1958, pp. 115–35.

Chapter 12 Background knowledge

1 William James, *Human Immortality*, Dent, London, 1917 (USA 1908), p. 51.
2 William Empson, 'Earth has Shrunk in the Wash', *Collected Poems*, Chatto & Windus, London, 1953, p. 28.
3 'Your Teeth are Ivory Towers', ibid., p. 46.

4 Louis MacNeice, *Ten Burnt Offerings*, Faber & Faber, London, 1952, pp. 37–43.
5 D. J. Enright, *Addictions*, Chatto & Windus, London, 1962, pp. 18–19.
6 Richard Burton, *The Anatomy of Melancholy*, I. iv. 1. Everyman edn, Dent, London, 1932, vol. I, p. 432. (This accessible edn modernizes the spelling and punctuation.)
7 Byron, *Childe Harold's Pilgrimage*, IV. lxii. Hannibal inflicted a shattering defeat on the Romans on the shores of the lake, known in Latin as *Trasimenus*, in 217 BC. Compare the first lines of Marlowe's *Dr Faustus*.
8 John Barth, *Giles Goat-Boy*, Secker & Warburg, London, and Penguin, Harmondsworth, 1967; Penguin edn, pp. 318–84.
9 Andrew Harvey, *Masks and Faces*, André Deutsch, London, 1978, pp. 5, 7, 11, 21, 22, 24–5, 26, 29–30, 37, 40, 43.
10 'Tarry, delight, so seldom met', A. E. Housman, *Collected Poems*, Cape, London, 1939, p. 119.
11 Sir John Davies, *Orchestra*, ll. 1–2.
12 *Hamlet*, IV. v. 42–3.
13 John Donne, *Holy Sonnets*, IX. ll. 10–11.
14 Robert Graves, *Collected Poems*, Cassell, London, 1975, pp. 149–50.
15 James Kirkup, *The Prodigal Son*, Oxford University Press, 1959, p. 24.
16 George Eliot, *Adam Bede*, ch. 8.
17 Wilfrid Gibson, 'The Plough', in *Collected Poems*, Macmillan, London, 1933, p. 372.
18 Louis MacNeice, 'Old Masters Abroad', in *Solstices*, Faber & Faber, London, 1961, p. 57.
19 John Donne, Elegie XIX. l. 27. *The Poems of John Donne*, ed. H. J. C. Grierson, Oxford University Press, 1933, p. 107.
20 *Paradise Lost*, Bk I, ll. 582–7.
21 *Muir's Historical Atlas*, ed. R. F. Treharne and H. Fullard, Philip, London, 6th edn, 1963.
22 George Herbert, 'Vertue'.

Chapter 13 Instruction and discussion

1 Thomas Middleton, *A Chaste Maid in Cheapside*, V. ii. 26–7.
2 W. K. Wimsatt, *The Verbal Icon*, University of Kentucky Press, Lexington, 1967.

Chapter 14 A few favourite fallacies

1 William Cowper, *The Task*, Bk VI, ll. 96–7.
2 Henry David Thoreau, *Walden*, Everyman edn, Dent, London, 1908, p. 91.
3 Patrick Quentin, *A Puzzle for Fools*, Gollancz, 1936, London, p. 14.
4 Byron, *Don Juan*, VIII, stanzas lxi–lxvii.
5 William Wordsworth, *The Prelude* (1805), IX, 71.
6 Thomas Hardy, 'De Profundis', *Poems of the Past and the Present*, Macmillan's pocket Hardy, London, 1921, p. 447.
7 John Milton, *Paradise Lost*, Bk IV, ll. 847–8.
8 Thomas Campbell, *Gertrude of Wyoming*, Bk III, ll. 1–2.
9 John Milton, *Samson Agonistes*, ll. 1648–9.

Chapter 15 Some rudiments of study technique

1 Dryden's trans. of Aulus Persius Flaccus, *Third Satyr*, ll. 16–21, 29–32.
2 Charles Lamb, *Last Essays of Elia*, Everyman edn, Dent, London, 1906, pp. 300–27.
3 Nick Perry, 'Second Best', *The Oxford Handbook*, Oxford University Students' Union, 1978, pp. 82–4.
4 Geoffrey Chaucer, *A Treatise on the Astrolabe*, *Complete Works*, ed. W. W. Skeat, Oxford University Press, 1912, p. 396.

Chapter 16 Essays and papers

1 Herman Melville, *Moby Dick*, World's Classics edn, Oxford University Press, 1920, p. 558.
2 Lucius Carie, 'Elegy on Dr Donne' in *The Poems of John Donne*, ed. H. J. C. Grierson, Oxford University Press, 1933, p. 349.
3 *Enchyridion*, Cent. IV, Cap. xcvii, p. 48, *Complete Works in Prose and Verse of Francis Quarles*, ed. A. B. Grosart; vol. 1, Edinburgh (privately printed), 1880.

Chapter 17 Examinations

1 *The Complete Poems of John Skelton*, ed. Philip Henderson, Dent, London, revised edn 1948, p. 164.

2 Joseph Morgan, *A Complete History of Algiers*, printed for the author by J. Bettenham, London, 1728; vol. 1, p. 274.
3 *Antony and Cleopatra*, V. ii. 186.
4 *King John*, III. i. 74.

For Product Safety Concerns and Information please contact our EU representative GPSR@taylorandfrancis.com
Taylor & Francis Verlag GmbH, Kaufingerstraße 24, 80331 München, Germany

www.ingramcontent.com/pod-product-compliance
Lightning Source LLC
LaVergne TN
LVHW011131300926
849160LV00008B/599
* 9 7 8 0 4 1 5 7 2 2 4 9 0 *